Natural Fibre Composites

Natural Fibre Composites

Manufacturing, Characterization, and Testing

By
Mohamed Zakriya G.
Ramakrishnan G.

CRC Press
Taylor & Francis Group
Boca Raton London New York

CRC Press is an imprint of the
Taylor & Francis Group, an **informa** business

First edition published 2021
by CRC Press
6000 Broken Sound Parkway NW, Suite 300, Boca Raton, FL 33487-2742

and by CRC Press
2 Park Square, Milton Park, Abingdon, Oxon, OX14 4RN

First issued in paperback 2022

ISBN 13: 978-0-367-55019-6 (pbk)
ISBN 13: 978-0-367-34589-1 (hbk)
ISBN 13: 978-0-429-32673-8 (ebk)

Typeset in Times
by Deanta Global Publishing Services, Chennai, India

Visit the companion website/eResources: https://www.routledge.com/Natural-Fiber-Composites-Manufacturing-Characterization-and-Testing/Zakriya-G-Govindan/p/book/9780367345891

Contents

Preface

The rapidly increasing consumption of petroleum-based products and its negative impact on the environment has led to an increase in environmental consciousness when it comes to sustainable materials and products. Natural fibres are categorized as an environmentally friendly material which have good properties compared to synthetic fibres. When it comes to the properties of the natural fibres, it is important to state that there are differences between one type of fibre and another. The performance of a composite depends on many factors such as structure, mechanical composition, physical properties, cell dimensions, and microfibrillar angle. Fibre-reinforced composites have received much attention based on different applications because of their good properties and the advantages they have been found to have over synthetic fibres.

The aim of this book is to highlight the fundamental aspects of the manufacturing, characterization, and testing of natural-fibre composites in such a manner that its contents are useful to readers in education, industry, or commerce. It thus fulfils the long-felt need for a comprehensive up-to-date textbook explaining this important sector of textile technology. Aspects covered include kinds of natural fibres and their special characteristics, interfacial compatibility, design of experiments towards manufacturing composites, modelling of natural fibre composites, process and production techniques of composites, product life cycle assessment and suitability, testing of composites, rheology and insulation behaviour of composites, applications of composites in engineering, applications of composites in artefact and furniture making, and recycling of natural fibre composites. The book is acceptable as a set text for textile courses from technician to degree and Master's level. It will also prove particularly suitable for professionals wishing to update or broaden their understanding of natural fibre composites. The contents have been arranged for the convenient use of different levels of readership with the text gradually progressing from an explanation of basic terminology and principles to eventually encompassing the most advanced aspects of the technology, including the application of concepts. The indexed and referenced format of the text is supplemented by labelled diagrams and photographs so that the book may also serve as a handy reference work for study and business purposes. Terminology is defined either according to Textile Institute terms and definitions or current usage in the industry and is supplemented as necessary by American or continental terminology. Internationally accepted methods of notation help to clarify explanations of composites.

It is particularly satisfying that this book has proved useful in education, industry, and commerce throughout the world. I hope the abovementioned additions will further increase its usefulness.

The authors wish to express their gratitude and heartfelt thanks to the management of KCG College of Technology, Chennai (A group of Hindustan University, Chennai) and the management of Kumaraguru College of Technology, Coimbatore (A group of Kumaraguru Institutions, Coimbatore) as well as friends and family members for their continuous support.

Lastly, we thank CRC Press/Taylor Francis Group for their generous cooperation at every stage of the book's production.

The authors are grateful to dedicate this book to the late A.W. Sahebjathi – a teacher (mother of Dr G. Mohammad Zakriya) – and Mr S. Govindan – a teacher (father of Dr G. Ramakrishnan).

Mohamed Zakriya G. and Ramakrishnan G.

Foreword

Composite materials play a significant role in engineering applications, and human-made fibres are the major components in them, which involve higher cost and are non-biodegradable. In contrast, natural fibres are cheap, abundantly available, non-toxic, and biodegradable, and can be treated chemically to enhance the mechanical properties, if required. Because of this, a tremendous surge in cutting technologies has emerged across the globe through extensive research on natural fibre-based composites.

The book titled *Natural Fibre Composites: Manufacturing, Characterization and Testing*, authored by Mohamed Zakriya G. and Ramakrishnan G., is a truly timed, authentic, and in-depth documentation of manufacturing, characterization, and testing of natural fibre composites, contributing to the enhancement of knowledge in this field. This book covers the appropriate use of accessible natural fibres towards the requirement and compatibility for industrial sustainability and upholds the natural characteristics of composites through technology and techniques. The inherent qualities of natural fibres are discussed with the design of experiments to the necessity. The durability of composites subjected to environmental conditions, biodegradability, environmental issues, product life cycle assessment, and testing methods are elaborated. Micro and macro mechanical properties, as well as the functional use of natural fibre-reinforced composites, are discussed. An in-depth discussion on important and emerging areas of natural fibre composites like the modelling of natural fibre composites, life cycle assessment, suitability, rheology, insulation behaviour, and recycling, besides dealing with the fundamental and other essential aspects of composites, has been nicely included in this book, which makes it a potential need for researchers and under-graduate/post-graduate students and an authentic manual for fabricators of composites and industrialists.

The authors' long experience in various capacities in textile mills, institutes, and research organisations has been reflected in their write-up. It is expected that the book will do exceptionally well among the textile fraternity and will enable them to learn basic concepts too, ultimately leading to hassle-free and effective handling.

I wish a grand success of the book!

Dr. J. N. Chakraborty
Professor
NIT, Jalandhar, India

The book *Natural Fibre Composites: Manufacturing, Characterization and Testing*, authored by Dr. Mohamed Zakriya G. and Dr. Ramakrishnan G., is an authentic and in-depth documentation of manufacturing, characterization, and testing of natural fibre-reinforced composites. Composite materials play a major role in many engineering applications. Natural fibres are cheap and abundantly available. Besides, they are nontoxic and can be treated physically or chemically in order to enhance the interfacial properties. Hence, many young scientists get the motivation to pursue research work in the field of natural fibre-reinforced composites.

The authors have thoroughly dealt with the fundamental aspects of composites, along with some important and emerging areas of natural fibre composites like modelling of natural fibre composites, life cycle assessment, rheology and insulation behaviour of composites, recycling of natural fibre composites, etc. This book will be helpful to the textile faculty for classroom teaching, and it will also enable students to take up interdisciplinary researches in the area of natural fibre-reinforced composites.

The authors' long experience in various capacities in the industry, academic institutes, and research organisations has been reflected in their contribution. It is expected that the book receives wide acceptance among the textile fraternity and will help them learn the basic concepts of the subject covered.

I wish a grand success of the book!

Thanking you,

Dr. Abhijit Majumdar
Institute Chair Professor
Department of Textile & Fibre Engineering
Indian Institute of Technology, Delhi

Over the past five decades, polymer composite materials have become an important class of materials in diverse engineering applications. This is due to their lightweight and outstanding, specific mechanical properties. These polymer composite materials are reinforced with human-made fibres such as glass, aramid, and carbon fibres. Energy-intensive processes of making these fibres result in higher embodied energy and higher carbon footprint of polymer composites. There is a growing demand worldwide to find alternative reinforcement materials which are less energy intensive, have reduced carbon footprint, and are derived from renewable sources. Fibres derived from natural plant sources are gaining attention now. Natural fibre-reinforced composites have become the focus of attention by the young researchers as well as sustainability-conscious companies and businesses. Natural fibres are cheap and abundantly available. They are environmentally benign and can be treated chemically in order to enhance the mechanical properties.

The book titled *Natural Fibre Composites: Manufacturing, Characterization and Testing*, authored by Mohamed Zakriya G. and Ramakrishnan G., is an authentic and in-depth documentation of manufacturing, characterization, and testing of natural fibre composites, contributing to the enhancement of knowledge.

This book covers the appropriate use of accessible natural fibres towards the requirement and compatibility for industrial sustainability and upholds the natural characteristics of composites through technology and techniques. The inherent qualities of natural fibres are discussed with the design of experiments to the necessity. The durability of composites subjected to environmental conditions, biodegradability, environmental issues, product life cycle assessment, and testing methods are elaborated. Micro and macro mechanical properties, as well as the functional use of natural fibre-reinforced composites, are discussed. An in-depth discussion on important and emerging areas of natural fibre composites like the modelling of natural fibre composites, life cycle assessment, suitability, rheology, insulation behaviour, and recycling, besides dealing with the fundamental and other essential aspects of composites, has been nicely included in this book, which makes it a potential need for researchers and under-graduate/post-graduate students and a helpful manual for fabricators of composites and industrialists.

Prof. Seeram Ramakrishna
Director, Center for Nanofiber and Nanotechnology Professor, Department of Mechanical Engineering National University of Singapore

About the Authors

Mohamed Zakriya, G., M.Tech., M.B.A., Ph.D. (Assistant Professor, Selection Grade-I, Department of Fashion Technology, KCG College of Technology, Chennai, Tamil Nadu, India) is an effective academician, trainer, and researcher in the field of fashion, technical textiles, and industrial management. He completed his doctoral degree (full-time) in Technical Textiles at Anna University, Chennai. He has 11 years of teaching and research experience in the field of fashion and textile technology. His areas of specialization are Technical Textiles, Clothing Comfort, Apparel Costing, Design Collection and Portfolio Management, Pattern Engineering, Home Textiles, Quality Control and Testing, Fashion Merchandising, Industrial Engineering, and Entrepreneurship Development studies. He has presented and published around 12 papers in national and international peer-reviewed high-impact factor journals and conferences and has filed six patents in IPR, Chennai. He has served as a technical advisor for Tirupur-based industries related to aspects of clothing comfort and product diversification in knitting sectors.

Ramakrishnan, G., is presently Professor and Head of the Department of Fashion Technology, coordinator of KCT-TIFAC CORE, and coordinator at the Natural Fibre Research Center at Kumaraguru College of Technology (KCT), Coimbatore. He has 33 years of experience in both the textile industry and academia. He was the recipient of the Gold Medal for securing First Rank in the Anna University PG Degree Examinations in the year 2004. He received his doctoral degree (PhD in Textile Technology) from Anna University, Chennai, in 2010. He was the recipient of the 'Best Mentor' award at the 1st AICTE-ECI Chhatra Vishwakarma Awards, organized by the All India Council of Technical Education (AICTE) and the Engineering Council of India (ECI) in New Delhi in 2017. He was also the recipient of the 2016 Dr Radhakrishnan Award for Best Teacher from KCT. He has published 50 papers in international peer-reviewed journals, including the international monograph *Textile Progress*, published by the Textile Institute, UK. He has coauthored a book on *Control Systems in Textile Machines*, published by Woodhead Publishing Pvt. Ltd., India. He is currently a reviewer of international journals and has presented nearly 40 papers at national and international conferences. He served as joint organizing secretary for two international textile conferences organized jointly with Texas Tech University, USA, and he has coordinated many national-level seminars and workshops. Dr Ramakrishnan helped to establish the prestigious TIFAC Center of Relevance and Excellence project in textile technology and machinery, the first of its kind in India, funded by the Department of Science and Technology of the Government of India at the KCT campus. The centre is currently carrying out product development, testing, and consultancy for the textile and apparel industries. He is an approved supervisor of research scholars at Anna University, Chennai, and so far five research scholars have completed their research and three scholars are pursuing PhDs under his guidance. He has carried out sponsored research projects funded by DST-TIFAC CORE and DST-SEED and DST-IDP, DST-WTI of the Government of India. He was nominated for the NRDC National Societal Innovation Award of the Year 2019 for the innovation/invention 'Low Weight Modified Jacquard for Handloom Weavers'.

1 Kinds of Natural Fibres and Its Special Characteristics

1.1 INTRODUCTION

To stimulate interest in natural fibre and materials, the year 2009 is considered as the International Year of Natural Fibres – IYNF.[1] Natural fibres represent a substantial cultivated biomass that contributes towards the economy and supports the nation's environmental policy. The widespread use and accessibility of natural fibres can reduce pressure on forests and agriculture. The usage of diverse natural fibres as raw materials will help to retain the ecological balance of nature.[2] Therefore, an increase in the usage of composites reinforced with natural fibre has been attempted due to numerous advantages such as easy availability, light weight, nonabrasive nature, low cost, low CO_2 emission, recyclable, renewability, and biodegradability.

Natural fibres from stalks, leaves, bast, stems, fruit, and seed plants possess individual and inherent physical, mechanical, and unique surface properties. Consequently, a vast knowledge of the special characteristics of natural fibres and their blends is required to manufacture quality composites. The properties of fibre together with a great effort in terms of their processing technologies to select a set of machineries and machine parameters and techniques for blending natural fibres will result in a compatible product at minimum cost.

The hydrophilic and hydrophobic nature of natural fibres results in distinctive and interesting sorption phenomena. Composites with natural fibres have advantages such as lower density, better thermal and sound insulation, and electrical insulation along with good mechanical properties, which favour the functional application of composites.

Composites with natural fibres are one of the smart alternative solutions to the problem of the negative environmental impact of industries. The annual yield of natural fibres in India is approximately 14.5 million tons compared to the world yield of 45.5 million tons.[3] Banana, coir, jute, sugar cane, turmeric, kapok, kenaf, palmyra, and pineapple are dominant. A tropical agricultural environment has huge potential for the growth and use of fibre derived from agricultural waste. Natural fibres are thus an important by-product of the extraction process and can be used as reinforcement material in composite products.[4]

Most natural fibres have considerable differences in mechanical properties and moisture absorption and have poor thermal characteristics, creating problems in the production of natural fibre composites. Fibre matrix adhesion has also become a topic of interest in research into producing good quality composites. For natural fibre composites to meet market requirements, it is becoming essential to consider the issues of fibre blend proportions, fibre matrix, process and method of selection, and product optimization through statistical methods.

Increasing demand for automotive materials with sound insulation qualities and low weight (for fuel efficiency) could be met by using natural fibres as alternatives. Natural fibres from renewable resources will offer a complete solution to the demands of industrial society and lead to the agricultural production of alternative materials. The use of natural plant and animal-based fibres as reinforced polymeric composites has become a key design criterion for designing and manufacturing elements for all industrial products.[5] Renewable agro crops, which are going to be incorporated to prepare or make sustainable composite materials, need to be considered for its life cycle assessment from its starting stage to the end stage at various phases including its recyclability and disposal.[6] Recovery of raw materials using gasification to methanol, biodegradation and CO_2 neutral thermal utilization is being examined.[7]

In automotive applications, the demand rate for natural fibre composites as alternatives to plastic composites was forecasted as 15%–20%. In building sectors, for some selected applications, it was forecasted to be 50%. Other consumer applications like flowerpots, tiles, and home and commercial furniture are emerging markets.[8]

Manufacturers are encouraged to produce composite materials to support environmentally friendly approaches such as (i) reducing the amount of toxic waste produced in manufacturing, (ii) reusing raw materials, (iii) repairing damaged or defected products, and (iv) recycling needs to be done up to the extent level, and finally (v) the scarabs are disposed of to the landfills.[5]

1.2   CLASSIFICATION OF NATURAL FIBRES

The use of natural fibres from plant, animal, and other sources as reinforcing or laminating material in the making of composite reduces concerns about environmental sustainability. A lot of natural fibres have a hollow space at their core called the *lumen*; this is made up of randomly arranged nodes that divide the fibre into distinct cells. The surface of natural fibres is rough, and these surface irregularities create good bonds to the matrix in a composite structure. The mechanical properties specific to individual natural fibres are important criteria in determining the end uses of composites. Tenacity and elongation up to the breaking point of natural fibres, especially flax, hemp, and ramie fibres, compete with E-glass fibres. The tensile strength and Young's modulus of the fibres increase with an increase percentage of cellulose.[9] As compared to synthetic fibres such as glass and aramid, natural fibres have low density, making them a suitable alternative for low weight applications where the weight of the material is a major problem.[10]

In the early 1990s, fourth-generation composites were made from fibre reinforcement of polymer to produce hybrid composites.[11] The presence of lignin content in natural fibres would be used to develop sustainable multifunctional composites along with matrix polymer strength. The stiffness of natural fibres depends on the percentage of cellulose contents and the arrangement of microfibrils.[12] A fibre's impact strength and load-bearing strength depend solely upon the alignment of the micrifibrils.[13,14] Fibres act as a load-bearing element, and the lignin, as a matrix load transfer medium, holds the fibre together and allows it to be oriented in a desired direction.[15] Agro-based bast fibres such as jute and flax are generally preferred by the composite industries because of their good structural and reinforcement performance. The cellulose fibrils in bast fibres hold the lignin and hemicellulose together.[16] The classification of natural fibres is shown in Figure 1.1.

1.3 CHARACTERIZATION OF FIBRES

1.3.1 PROTEIN FIBRE

In protein fibres, numerous reactive functional groups are present, with amino acids interconnected by peptide bonds. Protein contents are oriented parallel to the fibre axis. Protein fibre is commonly known for its lustre and softness. Alpaca goat fibre has good insulation properties. Angora fibre is very fine and warm to the touch. It is generally blended with wool fibre to reduce its cost. Camel hair has good sheen and insulation properties. Cashmere goat fibres feel soft and luxurious and have good absorbency.

Anti-parallel beta-sheet crystals form silk fibroins into high fibres with high tensile strength and toughness fibre.[17–19] The low-density silk fibres inherently possess good elongation and flame-resistant properties. Its mechanical properties are better than those of plant fibres.[20–23] Honey-bee silk fibre has good toughness and a stretchability of more than 200%.[24] Silk fibre blending is an efficient way to produce bio materials exhibits the combination features from its components.[25] Chicken-feather fibre shows good flexural modulus and noise reduction coefficient at its highest contribution percentage in the production of composite materials.[26–28] Human hair is also a protein fibre; its core constituent is keratin, and it is tough, intricate, and incredibly strong. A single strand of hair can bear a load of 100–150 grams.[29]

1.3.2 MINERAL FIBRE

Asbestos occurs naturally as fibre. A synthetic mineral fibre known as *rock wool* or *slag wool* is produced by blowing air or steam through molten rock or slag. It is soft and flexible and good insulator of electricity, heat, and corrosion. Mineral fibres are used as fillers in fireproofing and thermal insulation materials. Basalt is naturally available worldwide. It is eco-friendly in nature; its fibres are produced by the process of drawing and winding fibres from the melt. It has good fire resistance, is chemically inert and can tolerate impact load.[30]

Brucite is the mineral form of magnesium hydroxide; it has good anti alkaline properties. It is more stable in an alkaline medium than glass fibre. The moderate strength of this fibre can reach up to 900 MPa.[31]

1.3.3 CELLULOSE FIBRE

Leaf, bast, seed, fruit, cane, grass, reed, and stalk fibres have high cellulose content. The content of renewable cellulose fibre and its percentage weight of various fibres are shown in Table 1.1.

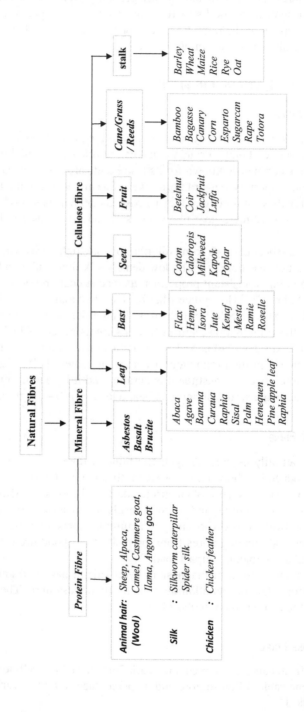

FIGURE 1.1 Classification of natural fibres.

TABLE 1.1

Cellulose content in various kinds of natural fibres[32-37]

Fibre	Cellulose in weight %
Abaca	56–63
Bagasse	33
Bamboo	73.8
Banana	61.5
Betelnut	35–64.8
Corn	15–20
Cotton	90
Coniferous wood	39–45
Coir	43
Curaua	71–74
Flax	82
Hemp	77.07
Henequen	77.6
Isora	74
Jute	63.24
Kenaf	65.7
Pineapple	71.6
Ramie	91
Oil palm	19
Sisal	66–72

Cellulose-based fibres are subjected to a mercerization process, which leads to fibrillation, in which the fibre bundle is broken into smaller fibres, in turn decreasing the fibre diameter. As part of this process, the roughness of the surface topography increases in line with increases in the aspect ratio and mechanical properties of composites.[38-41] Reaction sites can be improved in high cellulose content fibres by alkali treatment, which increases the mechanical bonding of fibres during the composite manufacturing process.[42,43] The fraction of the volume of cellulose and the degree of cellulose crystallinity determine the fibre's longitudinal Young's modulus.[44] After alkali treatment, cellulose fibre shows more aligned microfibrillar angle and thus has increased the load-bearing capacity.[45] The effectiveness of industrial composite material depends upon the fibre type and its cellulose content; adding a small quantity of matrix results in considerable changes in its physical, chemical and mechanical properties (e.g. tensile strength, Young's modulus).[46,47]

1.4 PERFORMANCE CRITERIA OF INDIVIDUAL NATURAL FIBRES IN MAKING COMPOSITES

The different structural features of Bombyx mori silkworm fibre at different lengths allow its properties to be tailored for optimal use. The arrangement of the hierarchical molecular structure provides resistance to damage and a tolerance limit in the bio-material development process.[48]

The presence of waxes and greases in wool fibres means that they are hydrophobic, and the fibre's scale-like structure with large pores helps to form capillary bridges. It is especially useful for oil spills. Human hair contains long chains of amino acids of a polymer called *keratin*. It consists of 50.65% carbon, 17.4% nitrogen, 5.0% sulphur, and 20.85% of Oxygen. It also contains low levels of glutamine, serine, cytosine, leucine, arginine, and valine. New kinds of composites using human hair with bio-polymers and natural fibres may find technical applications.[49]

Due to its high crystalline structure, chicken feather fibres (CFF) are biodegradable, durable, and stable. Composites made from blending CFF with boron oxide have an ideal load capacity and flame-retardant properties. Other avenues of research are focusing on combining CFF with polymer blends. Reports have shown that the flexural modulus of composites improves with the increment percentage of CFF. The scale-like structures of CFF made them very suitable for making thermal insulated composites.

Gomuti or Arenga pinnata fibre is traditionally used in Indonesia to make things such as brooms, water filters, and roof coverings; the characteristics of gomuti fibre are similar to those of coir fibre.[51]

Flax fibre has good tensile strength, moisture retention properties, and UV blocking ability, and the advantages of its low density and zero static charge offer many opportunities for the development of technical composite materials. For indoor use, it also has additional advantages of natural resistance to bacteria and insects.

The low impact resistance, higher frictional properties, bleachability and dyeing qualities, limited elongation, and rigidity of Jute fibre, along with its void content, mean that it offers acceptable interfacial bonding in composite manufacturing. Its stable thermal degradation and high load-bearing capacity are valuable features. It is applicable in civil engineering and automotive applications, the footwear and leather industry, and the building industry sectors. Higher-end material development research into this fibre is still ongoing.[53,54] Some of the physical properties of this fibre are given in Table 1.2. The natural composite manufacturing process is chosen based on the fibre's density, moisture retention, elongation, and Young's modulus value.

The hygroscopic nature of hemp fibre, its low CO_2 emissions, and its thinness make it effective in the development of composite materials for indoor use.[59] Sisal fibre, which is porous, is used to make materials for filtration and purification purposes.[60] The superior porosity of milkweed fibre makes it feasible for use in absorbing and holding oils in industry and the automobile sector.[61]

Pineapple fibre, with its high tensile strength, pliability, and good dyeing properties, means that is can be used as an alternative to leather.[62] Nanocellulose pineapple fibre is now used to make wound-dressing materials, implants and tissue engineering materials, and drug-delivery materials for use in the field of biotechnology.[63] Natural fibres have comparable mechanical properties to and are biocompatible with human tissue; moreover, natural fibres have no contrary effect on host tissue, which is a mandatory quality in materials that are to be used in biomedical applications.[64]

Presently, biodegradable natural fibres in plants are widely used as reinforcing elements in the development of polymer composites. Various aspects such as the

TABLE 1.2
Physical properties of natural fibres[55-58]

Fibre	Density (g/cm³)	Elongation (%)	Moisture regain (%)	Young's modulus (GPa)
Abaca	1.5	1.6	5.81	41
Bagasse	1.45	3–7	11	15–19
Bamboo	0.8–1.32	16	13.3	35.45
Banana	0.65–1.33	6.54	14.2–15.0	17.2–21.6
Betelnut	0.019–0.021	15–18	22–30	2.54–2.61
Corn	1.16	4.5	0.4–0.6	20.5
Cotton	1.54	3–10	8.5	6–10
Coir	0.67–1.15	1.17–2	10.5	4–13
Flax	1.27–1.55	3–4	12.4	27
Hemp	1.48	1.6	12	32
Henequen	1.2	6.1	10	24
Isora	1.35	6	9	18–20
Jute	1.46	1.8	13.75	20–25
Kenaf	0.15–0.55	3.4	1.3–5.5	23–27
Linen	1.4	2.7–3.5	10–12	50–70
Pineapple	1.25–1.60	–	13	50–71
Ramie	1.5–1.56	1.2–3.8	17.5	22
Oil palm	1.44	28	–	12
Sisal	1.33	4.3	11	17–22
Silk	1.52	20	8–11	6.3–11

plants' growth conditions, their maturity, and the harvesting methods used will have an effect on the properties of the natural fibre.

1.5 CONCLUSION

It is apparent that natural fibre–reinforced composites have become essential materials in recent years. They have been used in many industries such as in the aerospace, automotive, building, construction, and furniture industries. Among the benefits of naturally based composites are that they pose little or no health hazard, are low cost, lightweight, renewable, abundant, aesthetically attractive, and have good specific strength and stiffness. Natural fibre–reinforced composites can be considered as environmentally friendly materials compared to conventional fibre composites.

REFERENCES

1. www.fao.org
2. Asim, M., Khalina, A., Jawaid, M., et al. A review on pineapple leaves fibre and its composites. *International Journal of Polymer Science* 16, 950567, 2015.
3. Beg, M.D.H. and Pickering, K.L. Mechanical performance of kraft fiber reinforced polypropylene composites: Influence of fiber length, fiber beating and hydrothermal ageing. *Composites: Part A* 39, 1748–1755, 2008.

4. Venkatachalam, N., Navaneetha Krishnan, P., Rajsekar, R. and Shankar, S. Effect of pretreatment methods on properties of natural fiber composites: A review. *Polymers & Polymer Composites* 24(7), 555–566, 2016.
5. Cheung, H., Ho, M., Lau, K., Cardona, F. and Hui, D. Natural fibre-reinforced composites for bioengineering and environmental engineering applications. *Composites: Part B* 40, 655–663, 2009.
6. Annual Report. Environment Hong Kong 2005. Environmental Protection Department, The Government of the Hong Kong Special Administrative Region, 2005.
7. Riedel, U. and Nickel, J. *Applications of Natural Fiber Composites for Constructive Parts in Aerospace, Automobiles, and Other Areas.* https://application.wiley-vch.de/bo oks/biopoly/pdf_v10/vol10_16.pdf
8. Khondker, O.A., Ishiaku, U.S., Nakai, A. and Hamada, H. A novel processing technique for thermoplastic manufacturing of unidirectional composites reinforced with jute yarns. *Composites: Part A* 37(12), 2274–2284, 2006.
9. Pillai, C.K.S. Recent advances in biodegradable polymeric materials. *Material Science Technology* 30, 558–567, 2013.
10. Abhemanyu, P.C., Prassanth, E., Navin Kumar, T., Vidhyasagar, R., Prakash Marimuthua K. and Pramod, R. Characterization of natural fiber reinforced polymer composites. *AIP Conference Proceedings* 2080, 020005–1–020005-7, 2018. doi: 10.1063/1.5092888
11. Friedrich, K. and Jacobs, O. On wear synergism in hybrid composites. *Composites Science and Technology* 43, 71–84, 1992.
12. Thakur, V.K., Thakur, M.K., Ragavan, P. and Kessler, M.R. Progress in green polymer composites from lignin for multifunctional applications: A review. *ACS Sustainable Chemistry & Engineering* 2(5), 1072–1092, 2014.
13. Baillie, C. *Green Composites: Polymer Composites and the Environment.* North America: CRC Press, 2004.
14. Liese, W. Anatomy and properties of bamboo. Recent Research on Bamboos. In *Proceedings of the International Bamboo Workshop, Hangzhou, People's Republic of China, 6–14 October 1985.*
15. Abhemanyu, P.C., Prassanth, E., Navin Kumar, T., Vidhyasagar, R., Prakash Marimuthu, K. and Pramod, R. Characterization of natural fiber reinforced polymer composites. *AIP Conference Proceedings* 2080, 020005, 2019. doi: 10.1063/1.5092888
16. Zimniewska, M., Wladyka-rzybylak, M. and Manokowski, J. Cellulose fibres: Bio- and nano-polymer composites. In Kalia, S., Kaith, B.S. and Kaur, I. (eds.). *Bio and nano polymer composites.* Berlin, Germany: Springer, 2011, pp. 97–120.
17. Jin, H.J. and Kaplan, D.L. Mechanism of silk processing in insects and spiders. *Nature* 424, 1057–1061, 2003.
18. Hu, X., Shmelev, Karen, Sun, Lin, Gil, Eun-Seok, Park, Sang-Hyug, Cebe, Peggy and Kaplan, David L. Regulation of silk material structure by temperature-controlled water vapor annealing. *Biomacromolecules* 12, 1686–1696, 2011.
19. Omenetto, F.G. and Kaplan, D.L. New opportunities for an ancient material. *Science* 329, 528–531, 2010.
20. Pickering, K.L., Aruan Efendy, M.G. and Le, T.M. A review of recent developments in natural fibre composites and their mechanical performance. *Composites Part A: Applied Science and Manufacturing* 83, 98–112, 2016.
21. Ataollahi, S., Taher, S.T., Eshkoor, R.A., Ariffin, A.K. and Azhari, C.H. Energy absorption and failure response of silk/epoxy composite square tubes: Experimental. *Composites Part B Engineering* 43, 542–548, 2012.
22. Ude, A.U., Ariffin, A.K. and Azhari, C.H. Impact damage characteristics in reinforced woven natural silk/epoxy composite face-sheet and sandwich foam, coremat

and honeycomb materials. *International Journal of Impact Engineering* 58, 31–38, 2013.

23. Saba, N., Jawaid, M., Alothman, O.Y., Paridah, M.T. and Hassan, A. Recent advances in epoxy resin, natural fiber–reinforced epoxy composites and their applications. *Journal of Reinforced Plastics and Composites* 35, 447–470, 2016.

24. Sutherland, Tara D., Church, Jeffrey S. Hu, Xiao, Huson, Mickey G., Kaplan, David L. and Weisman, Sarah Single honeybee silk protein mimics properties of multi-protein silk. *PLoS One* 6, e16489, 2011. doi: 10.1371/journal.pone.0016489

25. Hu, X., Cebe, P., Weiss, A.S., Omenetto, F. and Kaplan, D.L. Protein-based composite materials. *Materials Today* 15(5), 208–215, 2012.

26. Barone, J.R. and Schmidt, W.F. Polyethylene reinforced with keratin fibers obtained from chicken feathers. *Composite Science and Technology* 65, 173–181, 2005.

27. Cheng, S., Lau, K., Liu, T., Zhao, Y., Lam, P.M. and Yin, Y. Mechanical and thermal properties of chicken feather fiber/PLA green composites. *Composites Part B* 40, 650–654, 2009.

28. Huda, S. and Yang, Y. Composites from ground chicken quill and polypropylene. *Composites Science and Technology* 68, 790–798, 2008.

29. Verma, A., Singh, V.K., Verma, S.K. and Sharma, A. Human hair: A biodegradable composite fiber – A review. *International Journal of Waste Resources* 6, 2, 2016.

30. Mingchao, W., Zuoguang, Z., Yubin, L., Min, L. and Zhijie, S. Chemical durability and mechanical properties of alkali-proof basalt fiber and its reinforced epoxy composites. *Journal of Reinforced Plastics and Composites* 27, 393–407, 2008.

31. Ma, X,. Ni, F.J. and Gu, X. Kindly replace the following reference: Identify characteristics of the brucite. *Journal of Highway and Transportation Research and Development* 23(1), 24, 2006. http://www.yataifr.com/en/hyzx/2020-04-14/56.html

32. Kumar, R., Obrai, S., Fengel, D., Mishra, S., Pothan, L.A., Jacob, M.D. and Sharma, A. Chemical modifications of natural fiber for composite material. *Pelagia Research Library* 2(4), 219–228, 2011.

33. Cook, J.G. *Handbook of Textile Fibre and Natural Fibres* (4th ed.). Watford, UK: Morrow Publishing, 1968.

34. Ramachandra, T.V., Kamakshi, G. and Shruthi, B.V. Bioresource status in Karnataka. *Renewable and Sustainable Energy Review* 8, 1–47, 2004.

35. Olesen, P.O. and Plackett, D.V. Perspectives on the performance of natural plant fibres. *Natural Fibres Performance Forum*. Copenhagen, Denmark, Forum Publication, 1999.

36. Paul, A., Joseph, K. and Thomas, S. Effect of surface treatments on the electrical properties of low-density polyethylene composites reinforced with short sisal fibres. *Composites Science and Technology* 57, 67–79, 1997.

37. Leao, A.L., et al. Fibra de Curauá: uma alternativa na produção de termoplá sticos reforçados. *Plástico Industrial* 3(31), 214–229, 2001.

38. Valadez-Gonzalez, A., Cervantes-Uc, J.M., Olayo, R. and Herrera-Franco, P.J. Effect of fiber surface treatment on the fiber-matrix bond strength of natural fiber reinforced composites.*Composites Part B: Engineering* 30(3), 309–320, 1999.

39. Kalia, S., Kaith, B.S. and Kaur, I. Pretreatments of natural fibers and their application as reinforcing material in polymer composites—A review. *Polymer Engineering and Science* 49(7), 1253–1272, 2009.

40. Pothan, L.A., George, J. and Thomas, S. Effect of fiber surface treatments on the fiber–matrix interaction in banana fiber reinforced polyester composites. *Composite Interfaces* 9(4), 335–353, 2002.

41. Oladele, I.O., Omotoyinbo, J.A. and Adewara, J.O.T. Investigating the effect of chemical treatment on the constituents and tensile properties of sisal fibre. *Journal of Minerals & Materials Characterization & Engineering* 9(6), 569–582, 2010.

42. Kalia, S., Kaith, B.S. and Kaur, I. *Cellulose Fibers: Bio- and Nano-Polymer Composites.* Heidelberg, Germany: Springer, 2011.
43. Cazaurang, M., Hirrera, P., Gonzaliz, I. and Anguilar, V.M. Physical and mechanical properties of henequen fibers. *Journal of Applied Polymer Science* 43(4), 749–756, 1991.
44. Xiong, X., Shen, S.Z., Hua, L., Liu, J.Z., Li, X., Wan, X. and Miao, M. Finite element models of natural fibers and their composites: A review. *Journal of Reinforced Plastics and Composites* 37(9), 617–635, 2018.
45. Osorio, L., Trujillo, E., Lens, F., Ivens, J., Verpoest, I. and Van Vuure, A.W. In-depth study of the microstructure of bamboo fibres and their relation to the mechanical properties. *Journal of Reinforced Plastics and Composites* 37(17), 1099–1113, 2018.
46. Ho, M.P., Wang, H., Lee, J.H., Ho, C.K., Lau, K.T., Leng, J. and Hui, D. Critical factors on manufacturing processes of natural fibre composites. *Composites Part B* 43, 3549–3562, 2012.
47. Matuana, L.M. and Balatinecz, J.J. Surface characterization of esterified cellulosic fibers by XPS and FTIR spectroscopy. *Wood Science and Technology* 35, 191–201, 2001.
48. Viney, C. *Natural Protein Fibers, Encyclopedia of Materials: Science and Technology* (2nd ed.), Oxford: Elsevier, 2001, pp. 5948–5956.
49. Popescu, C. and Hocker, H. Hair – the most sophisticated biological composite material. *Chemical Society Reviews* 37, 1282–1291, 2007.
50. Kuru, D., Akpinar Borazan, A. and Guru, M. Effect of chicken feather and boron compounds as filler on mechanical and flame retardancy properties of polymer composite materials. *Waste Management & Research* 36(11), 1029–1036, 2018.
51. Ticoalu, A., Aravinthan, T. and Cardona, F. A study into the characteristics of gomuti (*Arenga pinnata*) fibre for usage as natural fibre composites. *Journal of Reinforced Plastics and Composites* 33(2), 179–192, 2014.
52. Kers, J.P., Peetsalu, M., Saarna, A., Viikma, A., Krumme, A. and Menind, A. Prelimaray investigation into tensile characteristics of long flax fibre reinforced composites material. *Agronomy Research* 8(S1), 107–114, 2009.
53. Boubekeur, B., Belhaneche-Bensemra, N. and Massardier, V. Valorization of waste jute fibers in developing low-density polyethylene/poly lactic acid bio-based composites. *Journal of Reinforced Plastics and Composites* 34(8), 649–661, 2015.
54. Maity, S., Gon, D.P., Paul, P. and Singha, M. A review on jute nonwovens: Manufacturing, properties and applications. *International Journal of Textile Science* 1(5), 36–43, 2012.
55. Raghuveer, H., Desai, L., Krishnamurthy, T and Shridhar, N. Effectiveness of Areca (Betel) fiber as a reinforcing material in eco-friendly composites: A review. *Indian Journal of Advances in Chemical Science* S1, 27–33, 2016.
56. Dungani, R., Karina, M., Subyakto, A.S., Hermawan, D. and Hadiyane, A. Agricultural waste fibers towards sustainability and advanced utilization: A review. *Asian Journal of Plant Sciences* 15, 42–55, 2016.
57. Silva, R.V. and Aquino, E.M.F. Curaua fiber: A new alternative to polymeric composites. *Journal of Reinforced Plastics and Composites* 27(1), 103–112, 2008.
58. Kalita, B.B., Gogoi, N. and Kalita, S. Properties of Ramie and its blends. *International Journal of Engineering Research and General Science* 1(2), 1–6, 2013.
59. Kymäläinen, H.R., Koivula, M., Kuisma, R., Sjoberg, A.M. and Pehkonen, A., Technologically indicative properties of straw fractions of flax, linseed (*Linum usitatissimum* L.) and fibre hemp (*Cannabis sativa* L.) *Bioresource Technology* 94, 57–63, 2004.
60. Sahu, P. and Gupta, M.K. Sisal (*Agave sisalana*) fibre and its polymer-based composites: A review on current developments. *Journal of Reinforced Plastics and Composites* 36(24), 1759–1780, 2017.

61. Hubbe, M.A., Rojas, O.J., Fingas, M., Gupta, B.S. Cellulosic substrates for removal of pollutants from aqueous systems: A review. 3. Spilled oil and emulsified organic liquids. *BioResources* 8(2), 3038–3097, 2013.

62. Olivo, T. *Piñatex is a Sustainable, Leather Alternative.* Montvale, NJ, USA: Rodman Media Corp., Nonwoven Industry, 2015.

63. Cheiran, B., Leao, A., Souza, D., Thomas, S. and Pothan, L. Isolation of nano cellulose from pineapple leaf fibres by steam explosion. *Carbohydrate Polymers* 81(3), 720–725, 2010.

64. Janssen, J. *Designing and Building with Bamboo.* Eindhoven, the Netherlands: Technical University of Eindhoven, 2000.

2 Interfacial Compatibility

2.1 INTRODUCTION

Natural fibres are usually hydrophilic in nature and also regulate moisture phenomena. Their moisture content can reach from 8% to 12.6% and their degrading temperature is almost 180°–200°C.[1,2] The development of natural fibre–reinforced composites has increased the extensive array of applications for which they are suited. One of the most important characteristics of a fibre is its mechanical enactment due to interfacial bonding between the fibre and the polymer matrix.[2] Interfacial compatibility can be achieved by ensuring physical and chemical compatibility between the fibre and matrix. If this is not done, it leads to poor dispersion and adhesion and reduces the overall quality of the composite. This can be rectified through strategy modifications. Mechanical interlocking, electrostatic adhesion, interdiffusion, and chemical reactions are generally accountable for the interfacial bonding of natural fibre–reinforced composites; structural property associations can be identified by conducting indirect and direct interfacial assessments such as the microbond test (MT), single fibre fragmentation test (SFFT), and single fibre pull-out test (SFPT).

Effective inherent and superior interfacial bonding and stress transfer throughout the interface can be created through physical treatments such as heat treatment, solvent extraction, and corona and plasma treatments. Physicochemical processes such as UV bombardment, γ-ray, and laser treatments improve interfacial bonding. Chemical modifications can also improve the compatibility and bonding between the lignocellulosic molecules and hydrocarbon-based polymers[3–5] by changing the surface tension and impregnating the fibres.[6]

Lee and Wang found that fibre alkalization and the addition of coupling agents can be used to increase interfacial compatibility. Biocomposites treated with an alkali solution and silane coupling agent show improvement in the modulus of elasticity, tensile strength, and moisture resistance.[5,7] Polylactic acid (PLA), chitin, chitosan, starch, cellulose, collagen, lignin, polyhydroxyalkanoate, soy-based resins, and natural rubber are fully biodegradable and sustainable biopolymers. Bio-based materials have become increasingly appealing to most researchers and engineers in recent times as they offer an eco-friendlier alternative to traditional materials.[8–10]

Azwa et al. added a 10 mm thick water-based acrylic clear electrometric coating to bamboo/polyester composites to provide an additional barrier to the moisture penetration. Additionally, it provides durable protection against microorganisms, chemical exposure, rotting, UV rays, and abrasion.[11]

2.2 BLENDS WITHIN NATURAL FIBRES

The key components of natural plant fibre include cellulose, hemicellulose, lignin, pectin, waxes, and other low-molecule substances. Cellulose is the major structural element found in the form of lean rod-like crystalline microfibrils, aligned along the length of fibre.[12]

It is a semi-crystalline polysaccharide, consisting of a linear chain of hundreds to thousands of β-(1-4)-glycosidic bonds associated with D-glucopyranose along with the existence of a large number of hydroxyl groups. Hemicellulose is a low-molecular-weight polysaccharide that functions as a cementing matrix amongst cellulose microfibrils, which are present along with cellulose in almost all plant cell walls. Cellulose is crystalline, strong, and resistant to hydrolysis, while hemicellulose has a random, amorphous structure with poor strength. Furthermore, it is hydrophilic and can be simply hydrolyzed by dilute acids and bases.[13,14]

Lignin is one class of complex hydrocarbon polymers – that is, cross-linked phenol polymers – and it gives a plant its rigidity. It is relatively hydrophobic and aromatic in nature. Pectin is a structural heteropolysaccharide enclosed in the primary cell walls of plants, and it gives plants their flexibility. Wax and water-soluble ingredients are used to protect the fibre and its surface. The distinctive chemical structure makes the natural plant fibre hydrophilic in nature.[15]

Weak interfacial adhesion formation may arise due to a fibre's hydrophobic nature, limited processing temperatures, non-polar poor moisture resistance, inferior fire resistance, the formation of hydrogen bonds within the fibre itself, the development of bundles in the fibre, uneven spreading in a polar matrix during complex processing, and inadequate wetting of fibre by the matrix.[16–18]

Interfacial adhesion can be improved with various physical treatments, and one method used to change the surface structural properties of fibre is electrical discharge. Hosting surface crosslinking and altering the surface energy or creating reactive free radicals, groups, and thereby stimulating the mechanical bonding to the matrix. Chemical modification can permanently alter the natural fibre cell walls by grafting polymers onto the fibres, crosslinking the fibre cell walls, or using coupling agents.[19]

Combining natural fibres with bio plastics to manufacture fully biodegradable composite materials has drawn attention for various multifunctional needs. Since, numerous polymers are biodegradable and retain antimicrobial and antioxidant stuff.[20] Composites of polylactic acid (PLA) and natural plant fibre comprising <30 wt% of fibre have shown increased tensile modulus and reduced tensile strength. Adopting various chemical methods of transforming the surface of cellulosic fibres, such as cyanoethylation, esterification, and acetylation along with coupling agents/ compatibilizers, improves interface adhesion between the PLA matrix and the natural fibres. Natural fibres such as wood, bamboo and its flour, kenaf fibre, flax fibre, jute fibre, banana fibre, *Grewia optia* and nettle fibre, nanocellulose fibre, and jute-lyocell fibre have shown good compatibility with PLA.[21–25]

Polyhydroxyalkanoates (PHAs) belong to the family of linear polyesters made in nature by bacterial fermentation of sugar or lipids as intracellular carbon and energy storage granules.[26] Their degrading temperature is 200–250°C. A composite made up of PHAs fibre along with natural plant fibre shows high glass transition temperature, higher thermal stability, and higher heat distortion temperature. Crystallinity also improves with fibril loading and the composite can be utilized for various structural applications.[26–29] The chief copolymers or homopolymers of PHAs are poly (3-hydroxybutyrate) (PHB), poly (3-hydroxybutyrate-co-3-hydroxyvalerate) (PHBV), poly (3-hydroxybutyrate-cohydroxyhexanoate), and poly (3-hydroxybutyrate-co-hydroxyoctanoate). Jute and lyocell fibres and bamboo micro-fabril fibres show good compatibility with a PHB matrix. Tea plant fibre, beer spent-grain fibre, and wood flour show good compatibility with PHA. Wood fibre, recycled cellulose fibre, and cellulose nanowhisker show good compatibility with PHBV.[30]

Starch can be converted to thermoplastic starch (TPS) in the existence of plasticizers (water, glycol, glycerine, sorbitol, etc.) under the pressure of high temperature and shear by forming hydrogen bonds with the starch.[31] Pre-gelatinized cassava starch with *Luffa* fibre, thermoplastic corn starch with bleached *E. urograndis* pulp, cellulose derivatives/starch blends with sisal fibre, thermoplastic rice starch with cotton fibre, thermoplastic corn starch with sugar-cane and banana fibres, thermoplastic cassava starch with cassava bagasse cellulose nanofibrils, Thermoplastic maize starch with wheat straw cellulose nanofibrils, thermoplastic cassava starch with jute and kapok fibres all show good compatibility.[31–35]

2.3 BLENDS WITH SYNTHETIC FIBRES

Physical treatments on natural plant fibre alter the surface structure properties of the fibres. UV radiation, gamma radiation, and corona and plasma treatments are the most used physical techniques used to surface energy of natural plant fibre.[36–38] In the gamma radiation process, energy deposits on the natural fibre surface and radicals produced on the cellulose chain by hydrogen and hydroxyl abstraction break out some C-C bonds and initiate the chain scission. Similarly, peroxide radicals are generated while matrix polymers are exposed in the presence of oxygen. Gamma radiation creates these active sites in both fibre and matrix, in effect producing better bonding between the filler and polymer matrix and ultimately improving the mechanical strength of the composite.[36–38]

In the corona discharge process, electrons accelerated on the surface of the plastic rupture the long chain, creating multiple open ends and forming free valences. Due to the oxygenation caused by the electrical discharge, new carbonyl groups with higher surface energy are created, improving the chemical connection between the plastic molecules and the matrix. This corona discharge does not alter the strength and appearance of the material. This treatment is widely applied in natural cellulosic fibre composites with polyolefin matrix.[39]

When a composite of basalt fibre and epoxy is exposed to gamma irradiation, The treatment induces scission of the polymer chain and oxidation on the surface and inside the resin matrix. Due to higher tensile strength from treated composite, it creates a stable and flexural performance and shows low amplitude attenuation, but its interlaminar shear strength improves.[36]

When jute fibre along with PE and PP composite is exposed to UV radiation, the tensile strength and bending strength of the composite increases with the raising of UV radiation up to a radiation dose of 50 (mJ/cm^2). Compared to the untreated composite, the treated composite shows an 18% improvement in tensile strength and a 20% improvement in bending strength.[40]

When composites of miscanthus fibre with PP and miscanthus fibre with PLA are exposed to corona treatment, surface oxidation and an etching effect are caused. As a result, interfacial compatibility between the fibre and matrices increases. The mechanical and thermal properties (decomposition temperature, glass transition temperature, stress at yield, and Young's modulus) of the treated composites were greatly enhanced due to the improved interaction between the elements.[41]

When a composite of flax fibre and polyester composite is exposed to a 300 W plasma treatment process, its interfacial adhesion is improved. As a result, there is a 34% increase in tensile strength, a 31% increase in flexural strength, a 66% increase in flexural modulus, and a 39% increase in interlaminar shear strength.[42] When jute fibre is subjected to a mercerization process, the cementing layer in the cellulose fibrils, consisting of low molecular fats, lignin, pectin, and hemicellulose, is removed, creating a cleaner and rougher jute fibre surface; this improves resin wetting and interfacial adhesion. The interfacial shear strength of the jute fibre–epoxy composite increases by up to 40% after the mercerization process.[43] A process of acetylation increases the tensile and flexural strength of flax fibre–PP composites by up to 18%.[44]

When flax fibre is subjected to treatment with benzoylation peroxide, the result is a uniform dispersion of fibres within the polymer matrix. Less agglomerated fibres in the presence of dissociated fibres in the matrix contribute to a more efficient stress transfer from the matrix to the fibres upon stress solicitation, resulting in a flax fibre–PE composite with superior physical and mechanical properties.[45]

When jute fibre is exposed to siloxane treatment, bifunctional siloxane molecules create a molecular link across the interface via a covalent bond with both the cellulose surface and polymer resin of the composite. The use of this treated jute fibre in polyester and epoxy composites increases the tensile strength, flexural strength, and interlaminar shear strength of the composites.[46] The use of jute and sisal fibre treated with malleated coupling agents treated in composites with HDPE matrix composites shows significant improvements of 38% in tensile strength, 45% in flexural strength, and 67% in impact strength. The γ relaxation peaks shifted to higher temperature regions after the treatment due to the segmental hold of the matrix chains at the fibre surface, indicating enhanced interfacial adhesion.[26,47]

2.3.1 INTERFACIAL BONDING MECHANISM

Figure 2.1 shows the mechanisms of adhesion; typically one of these plays a leading role. Interdiffusion occurs due to intimate intermolecular interfaces between the molecules of the fibre and the matrix due to Van der Waals forces or hydrogen bonding.[48] Adsorption and diffusion are the two stages included in the adhesion mechanism. In the first stage, the fibre and the matrix must develop an intimate connection via spreading and penetration. After good wetting follows, permanent adhesion is established via molecular attraction, for example, electrostatic, Van der Waals, and covalent.

Good wetting leads to interdiffusion of molecules in both fibre and matrix. The degree of diffusion depends upon the chemical compatibility of the two distinct elements and the permeability of the substrate. Electrostatic adhesion is known to

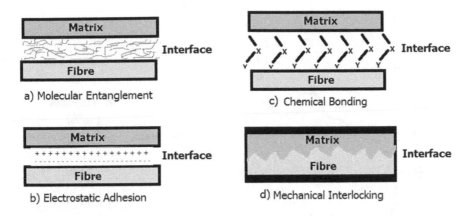

FIGURE 2.1 Interfacial bonding mechanisms.[48]

create anionic and cationic charges at the surfaces of the fibre and polymer matrix, accounting for the adhesion of the composite. Chemisorption arises between atomic and ionic bonds and creates chemical reactions in between.[48] Based on its surface chemistry, its chemical and physical bonds change in a process known as thermodynamic adhesion.

The matrix penetrates into valleys, holes, peaks, and other irregularities of the fibre substrate and thus mechanically locks into it.[48] Its sizes range from micron length scale to millimeter and diffusion entanglement at nanoscale in the fibre cell wall pores. Contact area adhesion, capillary forces, polymer flow least resistance, open pores structure of the natural fibre, cell wall shrinking and swelling are adhesion theories interrelated to mechanical interlocking of fibre and matrix.

Cell lumens and pits lead to micro penetration, while cracks induce macro penetration. The level of the penetration determines the quality of the bonding. The permeability of the matrix or resin varies in terms of its surface characteristics and by their direction for example, whether they are longitudinal, radial, or tangential. The etching of the fibre or polymer surface increases the surface roughness and in turn increases the contact area of fibre or polymer for adhesive penetration for mechanical interlocking.[49,50]

2.4 CHARACTERIZATION OF NATURAL FIBRE BLENDS

Research is now ongoing into the fabrication of biodegradable composites from natural low-cost raw materials.[51] As a result, renewable characters with thermoplastic nature materials are achieved through the plasticization of starch materials (TPS),[52] starch materials widely found in food grains (TPS).[52] The different processes associated with the process of the conversion of starch into TPS have been significantly improved due to research progress in the area of TPS.[53] The reinforcing of thermoplastic starch with natural lignocellulosic fibre in composite materials improves their mechanical enactment and eco-friendliness.[54]

The effect of fibre content on diffractogram patterns is shown in Figure 2.2. After accumulation of up to 20% of reinforcing fibres, no significant differences can be

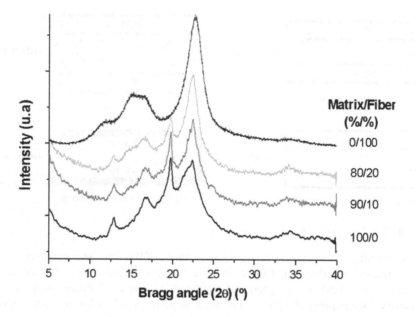

FIGURE 2.2   Diffractogram of corn starch TPS with sisal and hemp fibre.[33]

detected in the signals allotted to TPS. The depletion observed in Vh crystallinity peaks will be correlated with the decrease in the TPS content of the design. Additionally, fresh signals stare at $2\theta = 16.6°$ and $22.5°$, resulting in the crystalline share of cellulose in the fibres. Thermoplastic corn starch as matrix and fibre parts are indicated as X/Y percentage in the X-ray diffractogram.[33]

Changes take place in the share of crystallinity Vh of TPS after the merging of two reinforcing fibres. Autonomously, the type of fibre used and a gradual reduction in the crystallinity Vh of TPS is observed with the increase of fibre content. This reduction could be related to a decrease in the movement of starch molecules triggered by the presence of the firm reinforcing fibres, hampering the retrogradation of amylose chains. This hypothesis is strengthened by the outcomes of DMTA analysis, as shown in Figure 2.3.[33]

The results confirm that both tensile and flexural strength improves with the increased percentage of reinforcement fibres. Hemp strand composites provide better mechanical properties than those obtained with sisal. This is due to better fibrillation obtained when the TPS is mixed the hemp strands and indicates that mechanical anchoring is the main cause of the enhanced mechanical resistance.[33]

2.5   CHARACTERIZATION OF NATURAL FIBRE WITH SYNTHETIC FIBRE BLENDS

The use of natural fibres with synthetic fibre reinforced polymer composites have been used to enhance the performance of the composite material by reducing moisture absorption, thereby reducing negative environmental impacts and lowering the

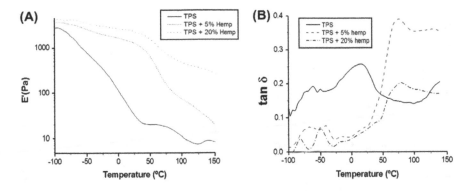

FIGURE 2.3 DMTA analysis of TPS of corn with hemp fibre. (A) E' vs. temp; (B) tan δ curves.[33]

consumption of energy level carbon footprints. Additionally, the cost of the composite material is comparable to that of glass fibre composite material, offering an alternative to chemical treatment by merging natural fibres with synthetic hydrophobic fibres.[55]

Akil et al. included glass fibres in a hybrid structure of polyester composites reinforced with jute fibres, resulting in the improvement of tensile and flexural properties. The addition of glass fibres into the jute polyester composite reduces its water absorption capacity.[56] In Figure 2.4, alkali-treated fibre hybrid composites show higher flexural and tensile modulus because of the improvement in interfacial compatibility.[57]

Researchers found that in hybridized woven flax–carbon and jute–carbon laminated composites, the escalation of carbon content led to an increase in the tensile properties. Higher percentages of jute fibres with carbon fibres make the composite material significantly stronger.[58,59] Based upon the reinforcement content, thermomechanical properties such as loss modulus (E″) storage modulus (E′), and tan δ damping factors of composite material vary.[60,61]

2.6 CONCLUSION

The need for natural fibre composites in engineering applications has progressively increased. Appreciable efforts have been made in terms of research to find the most appropriate and cost-effective coupling agents, surface modifiers, or dispersion aids for both the fibre and matrix to improve its interfacial bonds, depending on the end-use application. Nano-scale level interfacial characterization of the composite material helps to characterize the stress transfer, interaction, adhesion, and interfacial penetration of components present in composite materials. Furthermore, the use of nanotechnology in natural-fibre composites will necessarily play a significant role in the near future. The study of interfacial mechanisms and bonding in bio-based nanocomposites would greatly contribute to its development and will reduce the gap between scientific challenges and industrial production.

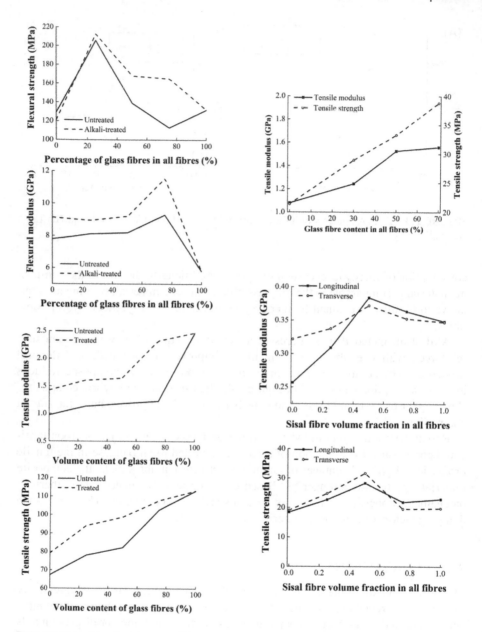

FIGURE 2.4 Interfacial compatibility of alkali-treated and untreated natural fibre.[57]

REFERENCES

1. Mohanty, A.K., Misra, M. and Hinrichsen, G. *Macromolecular Materials and Engineering* 276/277, 1, 2000.
2. Saheb, D.N. and Jog, J.P. Natural fiber polymer composites: A review. *Advances in Polymer Technology* 18(4), 351–363, 1999.

3. Zhou, Y., Fan, M. and Chen, L. Interface and bonding mechanisms of plant fibre composites: An overview. *Composites Part B* 101, 31–45, 2016. doi: 10.1016/j.compositesb.2016.06.055

4. Glasser, W.G., Taib, R., Jain, R.K. and Kander, R. Fiber-reinforced cellulosic thermoplastic composites. *Journal of Applied Polymer Science* 73(7), 1329–1340, 1999.

5. El-Abbassi, F., Assarar, M., Ayad, R. and Lamdouar, N. Effect of alkali treatment on Alfa fibre as reinforcement for polypropylene based eco-composites: Mechanical behaviour and water ageing. *Composite Structures* 133, 451–457, 2015.

6. Bledzky, A.K. and Gassan, J.M. Composites reinforced with cellulose based fibres. *Progress in Polymer Science* 24, 221, 1999.

7. Kang, J.T. and Kim, S.H. Improvement in the mechanical properties of polylactide and bamboo biocomposites by fiber surface modification. *Macromolecular Research* 19, 789–796, 2011.

8. Lee, S.-H. and Wang, S. Biodegradable polymers/bamboo fiber biocomposite with bio-based coupling agent. *Composites Part A: Applied Science and Manufacturing* 37, 80–91, 2006.

9. Methacanon, P., Weerawatsophon, U., Sumransin, N., Prahsarn, C. and Bergado, D.T. Properties and potential application of the selected natural fibers as limited life geotextiles. *Carbohydrate Polymers* 82, 1090–1096, 2010.

10. Dittenber, D.B. and Ganga Rao, H.V.S. Critical review of recent publications on use of natural composites in Azwa and Yousif, infrastructure. *Composites Part A: Applied Science and Manufacturing* 43, 1419–1429, 2012.

11. Azwa, Z.N. and Yousif, B.F. Physical and mechanical properties of bamboo fibre/polyester composites subjected to moisture and hygrothermal conditions. *Journal of Materials: Design and Applications* 233, 1065–1079, 2019. doi: 10.1177/1464420717704221

12. Azwa, Z.N., Yousif, B.F., Manalo, A.C. and Karunasena, W. A review on the degradability of polymeric composites based on natural fibres. *Materials and Design* 47, 424–442, 2013.

13. Wong, K.J, Yousif, B.F and Low, K.O. The effects of alkali treatment on the interfacial adhesion of bamboo fibres. *Proceedings of the Institution of Mechanical Engineers, Part L: Journal of Materials: Design and Applications* 224(3), 139–148, 2010.

14. Summerscales, J., Dissanayake, N.P.J., Virk, A.S. and Hall, W. A review of bast fibres and their composites. Part 1 – Fibres as reinforcements. *Composites Part A* 41(10), 1329–1335, 2010.

15. John, M.J. and Thomas, S. Biofibres and biocomposites. *Carbohydrate Polymers* 71(3), 343–364, 2008.

16. Araujo, J.R., Waldman, W.R. and De Paoli, M.A. Thermal properties of high density polyethylene composites with natural fibres: Coupling agent effect. *Polymer Degradation and Stability* 93(10), 1770–1775, 2008.

17. Dittenber, D.B. and Gangarao, H.V.S. Critical review of recent publications on use of natural composites in infrastructure. *Composites Part A: Applied Science and Manufacturing* 43(8), 1419–1429, 2012.

18. Dhakal, H.N., Zhang, Z.Y., Richardson, M.O.W. and Errajhi, O.A.Z. The low velocity impact response of non-woven hemp fibre reinforced unsaturated polyester composites. *Composite Structures* 81(4), 559–567, 2007.

19. Xie, Y., Hill, C.A.S., Xiao, Z., Militz, H. and Mai, C. Silane coupling agents used for natural fiber/polymer composites: A review. *Composites Part A* 41(7), 806–819, 2010.

20. Kovacevic, Z, Bischof, S, Fan, M. The influence of *Spartium junceum* L. fibres modified with montmorillonite nanoclay on the thermal properties of PLA biocomposites. *Composites Part B-Engineering* 78, 122–130, 2015.

21. Wang, Y., Weng, Y. and Wang, L. Characterization of interfacial compatibility of polylactic acid and bamboo flour (PLA/BF) in biocomposites. *Polymer Testing* 36, 119–125, 2014.

22. Masuelli, M.A.M. *Fiber Reinforced Polymers – The Technology Applied for Concrete Repair.* Rijeka: IntechOpen, 2013.
23. Sain, M., Suhara, P., Law, S. and Bouilloux, A. Interface modification and mechanical properties of natural fiber-polyolefin composite products. *Journal of Reinforced Plastics and Composites* 24(2), 121–130, 2005.
24. Mohanty, A.K., Misra, M. and Drzal, L.T. Surface modifications of natural fibers and performance of the resulting biocomposites: An overview. *Composite Interfaces* 8(5), 313–343, 2001.
25. Huda, M.S., Drzal, L.T., Misra, M. and Mohanty, A.K. Wood-fibre-reinforced poly (lactic acid) composites: Evaluation of the physicomechanical and morphological properties. *Journal of Applied Polymer Science* 102(5), 4856–4869, 2006.
26. Liu, D., Song, J., Anderson, D.P., Chang, P.R. and Hua, Y. Bamboo fiber and its reinforced composites: Structure and properties. *Cellulose* 19(5), 1449–1480, 2012.
27. Coats, E.R., Loge, F.J., Wolcott, M.P., Englund, K. and McDonald, A.G. Production of natural fiber reinforced thermoplastic composites through the use of polyhydroxybutyrate-rich biomass. *Bioresource Technology* 99(7), 2680–2686, 2008.
28. Bhardwaj, R., Mohanty, A.K., Drzal, L.T., Pourboghrat, F. and Misra, M. Renewable resource-based green composites from recycled cellulose fibre and poly(3-hydroxyb utyrate-co-3-hydroxyvalerate) bioplastic. *Biomacromolecules* 7(6), 2044–2051, 2006.
29. Krishnaprasad, R., Veena, N.R., Maria, H.J., Rajan, R., Skrifvars, M. and Joseph, K. Mechanical and thermal properties of bamboo microfibril reinforced polyhydroxybutyrate biocomposites. *Journal of Polymers and the Environment* 17(2), 109–114, 2009.
30. Srubar III, W.V., Pilla, S., Wright, Z.C., Ryan, C.A., Greene, J.P., Frank, C.W. and Billington, S.L. Mechanisms and impact of fiber–matrix compatibilization techniques on the material, characterization of PHBV/oak wood flour engineered biobased composites. *Composites Science and Technology* 72(6), 708–715, 2012.
31. Kaewtatip, K. and Thongmee, J. Studies on the structure and properties of thermoplastic starch/luffa fiber composites. *Materials and Design* 40, 314–318, 2012.
32. Curvelo, A.A.S., de Carvalho, A.J.F., Agnelli, J.A.M. Thermoplastic starch–cellulosic fibers composites: Preliminary results. *Carbohydrate Polymers* 45(2), 183–188, 2001.
33. Girones, J., Lopez, J.P., Mutje, P., Carvalho, A.J.F., Curvelo, A.A.S. and Vilaseca, F. Natural fiber reinforced thermoplastic starch composites obtained by melt processing. *Composites Science and Technology* 72(7), 858–863, 2012.
34. Alvarez, V. and Vazquez, A. Thermal degradation of cellulose derivatives/starch blends and sisal fibre biocomposites. *Polymer Degradation and Stability* 84(1), 13–21, 2004.
35. Prachayawarakorn, J., Sangnitidej, P. and Boonpasith, P. Properties of thermoplastic rice starch composites reinforced by cotton fiber or low-density polyethylene. *Carbohydrate Polymers* 81(2), 425–433, 2010.
36. Li, R., Gu, Y., Yang, Z., Li, M., Wang, S. and Zhang, Z. Effect of γ irradiation on the properties of basalt fiber reinforced epoxy resin matrix composite. *Journal of Nuclear Materials* 466, 100–107, 2015.
37. Zaman, H.U., Khan, M.A., Khan, R.A., Mollah, M.Z.I, Pervin, S. and Al-Mamun, M. A comparative study between gamma and UV radiation of jute fabrics/polypropylene composites: Effect of starch. *Journal of Reinforced Plastics and Composites* 29(13), 1930–1939, 2010.
38. Khan, M.A., Khan, R.A., Haydaruzzaman, Hossain, A. and Khan, A.H. Effect of gamma radiation on the physico-mechanical and electrical properties of jute fiber-reinforced polypropylene composites. *Journal of Reinforced Plastics and Composites* 28(13), 1651–1660, 2009.
39. Ragoubi, M., Bienaime, D., Molina, S., George, B. and Merlin, A. Impact of corona treated hemp fibres onto mechanical properties of polypropylene composites made thereof. *Industrial Crops & Products* 31(2), 344–349, 2010.

40. Zaman, H.U., Khan, M.A. and Khan, R.A. Improvement of mechanical properties of jute fibers – polyethylene/polypropylene composites: Effect of green dye and UV radiation. *Polymer-Plastics Technology and Engineering* 48(11), 1130–1138, 2009.
41. Ragoubi, M., George, B., Molina, S., Bienaime, D., Merlin, A., Hiver, J. and Dahoun, A. Effect of corona discharge treatment on mechanical and thermal properties of composites based on miscanthus fibres and polylactic acid or polypropylene matrix. *Composites Part A* 43(4), 675–685, 2012.
42. Sarikanat, M., Seki, Y., Sever, K., Bozaci, E., Demir, A. and Ozdogan, E. The effect of argon and air plasma treatment of flax fiber on mechanical properties of reinforced polyester composite. *Journal of Industrial Textiles* 45(6), 1252–1267, 2016.
43. Doan, T., Brodowsky, H. and Mader, E. Jute fibre/epoxy composites: Surface properties and interfacial adhesion. *Composites Science and Technology* 72(10), 1160–1166, 2012.
44. Bledzki, A.K., Mamun, A.A., Lucka-Gabor, M. and Gutowski, V.S. The effects of acetylation on properties of flax fibre and its polypropylene composites. *Express Polymer Letters* 2(6), 413–422, 2008.
45. Wang, B., Panigrahi, S., Tabil, L. and Crerar, W. Pre-treatment of flax fibers for use in rotationally molded biocomposites. *Journal of Reinforced Plastics and Composites* 26(5), 447–463, 2007.
46. Seki, Y. Innovative multifunctional siloxane treatment of jute fiber surface and its effect on the mechanical properties of jute/thermoset composites. *Materials Science and Engineering: A* 508(1–2), 247–252, 2009.
47. Mohanty, S., Verma, S.K. and Nayak, S.K. Dynamic mechanical and thermal properties of MAPE treated jute/HDPE composites. *Composites Science and Technology* 66(3–4), 538–547, 2006.
48. Mohanty, S. and Nayak, S.K. Interfacial, dynamic mechanical, and thermal fiber reinforced behavior of MAPE treated sisal fibre reinforced HDPE composites. *Journal of Applied Polymer Science* 102(4), 3306–3315, 2006.
49. Kim, J.K. and Pal, K. *Recent Advances in the Processing of Wood-Plastic Composites.* Berlin, Germany: Springer Science & Business Media, 2011.
50. Migneault, S., Koubaa, A., Perre, P. and Riedl, B. Effects of wood fiber surface chemistry on strength of wood-plastic composites. *Applied Surface Science* 343, 11–18, 2015.
51. Tran, P., Graiver, D. and Narayan, R. Biocomposites synthesized from chemically modified soy oil and biofibers. *Journal of Applied Polymer Science* 102, 69–75, 2006.
52. Raquez, J.M., Nabar, Y., Narayan, R. and Dubois, P. New developments in biodegradable starch-based nanocomposites. *International Polymer Processing* 22(5), 463–70, 2007.
53. Bogoeva-Gaceva, G., Avella, M., Malinconico, M., Buzarovska, A., Grozdanov, A., Gentile, G. and Errico, M.E. Natural fiber eco-composites. *Polymer Composites* 28(1), 98–107, 2007.
54. Van Soest, J.J.G. and Essers, P. Influence of amylose-amylopectin ratio on properties of extruded starch plastics sheets. *Pure and Applied Chemistry* A34, 1665–1689, 1997.
55. Safri, S.N.A., Sultan, M.T.H., Jawaid, M., Jayakrishna, K. Impact behaviour of hybrid composites for structural applications: A review. *Composites Part B: Engineering* 133, 112–121, 2017. doi: 10.1016/j.compositesb.2017.09.008
56. Akil, H.M., Santulli, C., Sarasini, F., Tirillo, J. and Valente, T. Environmental effects on the mechanical behavior of pultruded jute/glass fibre-reinforced polyester hybrid composites. *Composites Science and Technology* 94, 62–70, 2014. doi: 10.1016/j.compscitech.2014.01.017
57. Dong, C. Review of natural fibre-reinforced hybrid composites. *Journal of Reinforced Plastics and Composites* 37(5), 331–348, 2018. doi: 10.1177/0731684417745368
58. Kureemun, U., Ravandi, M., Tran, L.Q.N., Teo, W.S., Tay, T.E. and Lee, H.P. Effects of hybridization and hybrid fibre dispersion on the mechanical properties of woven

flax-carbon epoxy at low carbon fibre volume fractions. *Composites Part B: Engineering* 134, 28–38, 2018. doi: 10.1016/j.compositesb.2017.09.035

59. Ramana, M.V., Ramprasad, S. Experimental investigation on jute/carbon fibre reinforced epoxy based hybrid composites. *Materials Today: Proceedings* 4, 8654–8664, 2017. doi: 10.1016/j.matpr.2017.07.214

60. Essabir, H., Bensalah, M.O., Rodrigue, D., Bouhfid, R. and Qaiss, A. Structural, mechanical and thermal properties of bio-based hybrid composites from waste coir residues: Fibers and shell particles. *Mechanics of Materials* 93, 134–144, 2016. doi: 10.1016/j.mechmat.2015.10.018

61. Essabir, H., Boujmal, R., Bensalah, M.O., Rodrigue, D., Bouhfid, R. and el kacem Qaiss, A. Mechanical and thermal properties of hybrid composites: Oil-palm fiber/clay reinforced high density polyethylene. *Mechanics of Materials* 98, 36–43, 2016. doi: 10.1016/j.mechmat.2016.04.008

3 Design of Experiments towards Manufacturing of Composites

3.1 INTRODUCTION

Optimization is a process involving the search for optimal profiles to find one control variable or more than one variable in a time. Using experimental design methods, the number of experiments can be reduced by combining the variables and regulating synchronized studies and their effects. The design of experiments saves time and money and it is the most economical approach, and hence it is preferred by most researchers in all research domains.[1,2]

To achieve appropriate quality of composite products by design of experiments, Taguchi recommends a three-part designing sequence comprising system design, parameter design, and tolerance design. The conceptualization and synthesis of a product or process are considered in system design. Parameter design involves finding appropriate design factors to make the system less sensitive when subjected to uncontrollable noise variations. Tolerance design deals with customizing the products or process to reduce total manufacturing and lifetime costs.

The Taguchi method scans the consequences of the factors on the responses simultaneously by combining the factors, and it considerably reduces the number of experiments needed (Table 3.1).[4,5] Orthogonal arrays are used to study a large number of variables by doing a small number of experiments, the result being that research and development (R&D) costs are considerably reduced. Analysis of variance (ANOVA) is also used to analyze the impact of results as well as to determine error variance and its relative importance in various factors.[6]

3.2 EXPERIMENTS BASED ON SETTINGS OF FACTORS

Maryam Farbod, in his research work, adopted 04 levels of design process by regulating 06 factors of parameter.

Based on his 06 factor of parameter, 16 experiments were executed (Tables 3.2 and Table 3.3).

3.3 EXPERIMENTS BASED ON OUTPUTS

Zakriya et al., in their research work, prepared composites approximately 2.5–3.5 kg/m^2 weight of composite panel. By considering the range of weight of material, the number of experiments was decided by Box and Behnken's factorial design.

TABLE 3.1
Sample of orthogonal arrays in Taguchi's approach[3]

Factors				Factors								factors			
Run	A	B	C	Run	A	B	C	D	E	F	G	Run	A	B	C	D
1	1	1	1	1	1	1	1	1	1	1	1	1	1	1	1	1
2	1	2	2	2	1	1	1	2	2	2	2	2	1	2	2	2
3	2	1	2	3	1	2	2	1	1	2	2	3	1	3	3	3
4	2	2	2	4	1	2	2	2	2	1	1	4	2	1	2	3
				5	2	1	2	1	2	1	2	5	2	2	3	1
				6	2	1	2	2	1	2	1	6	2	3	1	2
				7	2	2	1	1	2	2	1	7	3	1	3	2
				8	2	2	1	2	1	1	2	8	3	2	1	3
												9	3	3	2	1

L_4 (2^3) array L_8 (2^7) array L_9 (3^4) array

TABLE 3.2
Influencing factors on research work[7]

Factor	Level 1	Level 2	Level 3	Level 4
Type of CNT (A)	Functionalized SWCFD	Functionalized MWCNT	SWCNT	MWCNT
Type of solvent (B)	Toluene	THF	Chloroform	–
Type of PS (C)	Type I	Type II	–	–
Film drying temperature (D)	25	60	–	–
CNT weight percentage (E)	0.5	1.0	1.5	–
Mixing time duration (F)	1 h	2 h	–	–

Note: SWCNT, single-walled carbon nanotubes; MWCNT, multi-walled carbon nanotubes; PS, polystyrene

Assessment of the responses on or after this factorial design model is appropriate only when the independent variables are within the range for which the model has been developed. Here, factor A and B are taken as dependent variables and factor C as an independent variable. According to the factors, the fabrication process was controlled on the composite material. Table 3.4 displays the coded and actual values of the three parameters measured, and the fifteen arrays of experimental combinations under which the composites were produced are given in Table 3.5.[8]

In this model, thermal resistance responses show a good significant p value of 0.09% compared to all other thermal responses. Table 3.6 shows the coefficient and constants of the response surface equations. A negative predicted R^2 implies that the overall mean is better in the current model response than in the predicted model. Knowing the predicted R^2 values helps to understand the efficiency of the current model in relation to the predicted model.

TABLE 3.3
Number of experiments and its combination[7]

Experiment run	A	B	C	D	E	F
1	1	1	1	1	1	1
2	2	2	1	1	2	1
3	1	3	2	2	3	1
4	2	1	2	2	1	1
5	2	3	2	1	1	2
6	1	1	2	1	2	2
7	2	1	1	2	3	1
8	1	2	1	2	1	2
9	1	1	1	2	1	3
10	2	3	1	2	2	3
11	1	2	2	1	3	3
12	2	1	2	1	1	3
13	2	2	2	2	1	4
14	1	1	2	2	2	4
15	2	1	1	1	3	3
16	1	3	1	1	1	4

TABLE 3.4
Design of composite using Box and Behnken model

Factor	Name	Unit	Minimum	Maximum	Coded values		Mean	Std. Dev.
A	Weight of jute lap	g/m²	1250	2450	−1 = 1250	1 = 2450	1850	438.17
B	Weight of HCP lap	g/m²	750	1750	−1 = 750	1 = 1750	1250	365.14
C	Thickness	mm	4	5	−1 = 4	1 = 5	4.5	0.36

By adopting design of experiments (DoE), experiment samples are decided, reducing the cost and time associated with the research and the number of samples involved. Optimizations through result analysis task are now simplified with this DoE.

3.4   CONCEPTUAL DESIGN

Conceptual design is an initial phase of the design process; it covers outlines of purposes and form of aesthetics, safety, legal requirements, efficiency and useful life, and so on. It comprises the design strategies, interactions, experiences, and processes. It also includes the logical process of developing research knowledge into a realistic and appropriate research design.

In general, international and national standards command many characteristics for a product; particularly in the case of products where community safety is involved. In such circumstances, conceptual design will be constrained, and the limitations of initial and complete design become imprecise. In principle, standards for structural

TABLE 3.5

Fifteen combinations of samples using coded values

	Weight of jute lap (A)	Weight of HCP lap (B)	Thickness (C)
Sample no.	(g/m^2)	(g/m^2)	(mm)
S1	1850	1750	5.0
S2	1850	1250	4.5
S3	1850	1750	4.0
S4	1250	750	4.5
S5	2450	1250	4.0
S6	2450	750	4.5
S7	2450	1750	4.5
S8	1850	750	4.0
S9	1250	1750	4.5
S10	2450	1250	5.0
S11	1850	750	5.0
S12	1250	1250	4.0
S13	1850	1250	4.5
S14	1250	1250	5.0
S15	1850	1250	4.5

TABLE 3.6

Response surface equation and coefficient of multiple correlations of composites

Response	Response surface equation in terms of coded factors	Coefficient of multiple correlations (R^2)	Predicted R^2
Thermal conductivity	0.022 − 0.001313A − 0.001475B − 0.0 003375C + 0.0037AB + 0.000375AC − 0.00005BC	0.7437	−0.2739
Thermal resistance	0.22 + 0.012A + 0.010B + 0.024C − 0.028AB + 0.002775AC + 0.0017BC − 0.011A2 − 0.013B2 + 0.0004667C2	0.9808	0.8238
Thermal transmittance	4.90 − 0.35A − 0.33B − 0.55C + 0.82AB − 0.022AC + 0.013BC	0.8385	0.3648
Thermal diffusivity	2.04 − 0.48A − 0.43B + 0.098C + 0.8 0AB + 0.17AC − 0.046BC	0.7656	−0.0043

A = weight of jute in grams, B = weight of HCP in grams, C = thickness of the composite in mm

lay-out calculations must comply with regulations. But for conceptual design aspects the numerical calculations are too difficult and too complex. Standards define the terms of agreements in guiding words. For conceptual designers, standards can be created at a later stage, even after that minimum usable prototype model will help to achieve their goals.[9]

Conceptual design specifies the geometry of structural elements, such as cylinders, plates, and beams, with approximate dimensions; shell width, breadth, and depth; failure criteria; and load requirements. A simplified design analysis based on the elements, determination, or selection of the shell wall material is carried out using analytical methods by using appropriate PC design programs or simplified finite element analysis. Acute regions in the design structure are identified and extreme loads and moments in the shell wall material are found; all this information is used as input data for classical laminate theory (CLT) models used to design composite laminates in the second phase. The effect of changing the number of plies, the layup or down, orientation angle and ply materials on shell wall properties are explored, if necessary by using optimization or iteration analysis.[9]

The third phase of the conceptual design process is about manufacturing the concept structure, and it may involve numerous idealized elements. The draping of composite plies finished by computer aided design (CAD), and the apparent geometry is modelled by draping codes; this determines any need to change the orientation of plies and fibres by mapping onto the composite structure surface. In the last phase, finite element models are created of the composite structure with true realistic ply and laminate layups; these will form the basis for a detailed structural analysis under design loads. If the composite structure does not meet or satisfy the design criteria, then the designer may modify the composite material, geometry or ply layup or choose alternatives, and further repeat the series until a successful design is attained.[9]

3.5 DESIGN OF SAMPLES

Rajkumar et al., in their research, experimented with using selvedge waste silk fibre to reinforce a polypropylene fibre matrix using a compression moulding process. Various compression moulding temperatures (165–185°C), times (7–15 min) and pressures (35–45 bar) were selected with respect to the mechanical attributes of the silk fibre–reinforced polypropylene composite. After analyzing the results, they found that the composite sample which was prepared at a temperature of 180°C temperature for 7 minutes at 35 bars of pressure had good mechanical properties. It also shows optimized responses to the predicted model. The Box–Behnken model Design (BBD) with three factors at three levels was used to design the samples (Table 3.7 and Table 3.8).[10]

TABLE 3.7
Design ranges and levels using BBD[10]

Variables	Range and level		
	−1	0	+1
Temperature in °C	165	175	185
Time in minutes	7	11	15
Pressure in bar	35	40	45

TABLE 3.8
Design of samples[10]

Run	Temperature (°C)	Time (min)	Pressure (Bar)
1	175	15	45
2	175	11	40
3	185	15	40
4	185	11	45
5	175	11	40
6	165	7	40
7	175	7	45
8	175	11	40
9	175	11	40
10	175	15	35
11	185	11	35
12	185	7	40
13	165	11	45
14	165	15	40
15	165	11	35
16	175	11	40
17	175	7	35

A similar kind of experiment was done by Mat Kandar et al., in which they optimized the hot-press forming process for woven flax–PLA composites using three independent process variables::moulding temperature, time, and pressure. Through BBD, a set of 16 experiments were established and run. The optimum results attained on the composites for the variables set for the compression moulding technique parameters were 3 min, 200°C, and 30 bar in order to achieve an impact strength 48.902 kJ/m^{-2}.[11]

3.6 CONCLUSION

Composites have been extensively used in structural and engineering applications. Attributes such as high strain to failure, stiffness and high specific strength, infinite shelf life, better impact strength, shorter processing cycle time and recyclability are in high demand. This demand can be met by selecting optimum process parameters initiated through DoE. Further, ANOVA is adopted to find the significant factors statistically, providing a clear picture of the role of the parameter and its response and the level of significance of the considered factor. The 'F' test is also carried out to find the significance of the process parameter. The high 'F' value denotes that the factor is highly significant to the response of the process. One of the most important purposes of experiments related to fabrication of composites is to achieve the desired mechanical properties with the optimal parameters.[12]

REFERENCES

1. Norberg, Ida, Nordström, Ylva, Drougge, Rickard, Gellerstedt, Göran and Sjöholm Elisabeth A new method for stabilizing softwood kraft lignin fibers for carbon fiber production. *Journal of Applied Polymer Science* 128(6), 188–198, 2013.
2. Kusworo, T. D., Ismail, A. F. and Mustafa, A. Experimental design and response surface modeling of PI/PES-zeolite 4a mixed matrix Membrane for CO_2 separation. *Journal of Engineering Science and Technology* 10, 1116–1130, 2015.
3. Wysk, R.A., Niebel, B.W., Cohen, P.H. and Simpson, T.W. *Manufacturing Processes: Integrated Product and Process Design.* New York: McGraw Hill, 2000.
4. Canel, T., Kaya, A.U. and Çelik, B. Parameter optimization of nanosecond laser for micro drilling on PVC by Taguchi method. *Optics & Laser Technology* 44, 2347–2353, 2012.
5. Sivarao, S., Milkey, K.R., Samsudin, A.R., Dubey, A.K. and Kidd, P. *Jordan Journal of Mechanical and Industrial Engineering* 8, 35–42, 2014.
6. Esmizadeh, E., Naderi, G., Ghoreishy, M.H.R. and Bakhshandeh, G.R. Optimal parameter design by Taguchi method for mechanical properties of NBR/PVC nanocomposites. *Iranian Polymer Journal* 20, 587–596, 2011.
7. Farbodi, M. Application of Taguchi method for optimizing of mechanical properties of polystyrene-carbon nanotube nanocomposite. *Polymers & Polymer Composites* 25(2), 177–184, 2017.
8. Zakriya, G.M., Ramakrishnan, G., Palanirajan, T. and Abinaya, D. Study of thermal properties of jute and hollow conjugated polyester fibre reinforced non-woven composite. *Journal of Industrial Textiles* 46(6), 1393–1411, 2017.
9. Kretsis, G. and Johnson, A.F. Conceptual design of composite structures. *Comprehensive Composite Materials II* 8, 26–45, 2018. doi: 10.1016/B978-0-12-803581-8.10049-9
10. Govindaraju, R. and Jagannathan, S. Optimization of mechanical properties of silk fiber-reinforced polypropylene composite using Box–Behnken experimental design. *Journal of Industrial Textiles* 47(5), 602–621, 2016. doi: 10.1177/1528083716667257
11. Mat Kandar, M.I. and Akil, H.M. Application of Design of Experiment (DoE) for parameters optimization in compression moulding for flax reinforced biocomposites. *Procedia Chemistry* 19, 433–440, 2016.
12. Suresh, S. and Senthil Kumar, V.S. Experimental determination of the mechanical behavior of glass fiber reinforced polypropylene composites. *Procedia Engineering* 97, 632–641, 2014.

4 Modelling of Natural Fibre Composites

4.1 INTRODUCTION

The finite element method is popular in the design engineering community. Through finite element analysis (FEA), a model of any composite structure with its complexity in shape, material selection, boundary conditions, load-withstanding ability, and other influencing factors can be analyzed under specific conditions. Virtual experiments could be done and viewed via a graphical user interface. Researchers carry out many iterations using FEA to optimize the outcomes so as to attain higher accuracy in the structure, enhance its lifetime, account for uncertainties, and reduce downtime in the product development process.[1]

FEA can be used to calculate the mechanical properties of discontinuous fibre composites and helps to determine the source of variability; in general, accuracy is imperfect owing to the quality of the fibre's structural design considered at the meso-scale level as a representative volume element (RVE). Currently, numerical models are used to generate RVE, which can be categorized into three main groups – sedimentation, hard, and soft models – along with their limitations. Sedimentation algorithms are tough systems that impose restrictions on the fibre alignment distribution. While the restriction on the ceiling volume fraction is condensed, sedimentation algorithms are computationally costly, and forming meshes for rigid models would be problematic, due to the small distances between the bundles of the rigid structures. In a hard model, it prevents bundle-to-bundle penetrations; fibre volume fractions are restricted due to fibre jamming, and this is usually experienced when the pockets of free space are too small to accept other inclusions. In soft models, overlapping of fibre bundles is permitted to occur, and hence it is unrealistic, as there is no restriction on the fibre volume fraction that is imposed. Allowing bundle-to-bundle penetration also creates incorrect load transfer paths at bundle crossovers.[2]

RVE size is linked to the fibre length and tow size and can be several orders of magnitude larger than the scale of the reinforcement. Computation time is one of the primary concerns in meso ranges discontinuous fibre composites materials. Two-dimensional (2D) models are a computationally inexpensive option, using a 1D linear beam element to represent the fibre bundles, randomly distributed in 2D space. This overlooks fibre crimping and allows bundle-to-bundle penetration, as all bundles are deposited on the same plane, reducing the accuracy. 2D models are also inclined to be over-stiff, as interconnecting bundles are rigidly bonded at the intersection points, increasing as the RVE thickness increases.[2] Three numerical algorithms – fibre kinematics, a custom Delaunay meshing algorithm, and tensile modulus predictions – were integrated to create a realistic simulation of the network of a composite structure. Fibre tortuosity was simulated using fibre bending and twist compaction of fibre kinematics. Compressive, tensile, and in-plane shear properties were simulated using a Delaunay meshing algorithm with respect to the RVE fibre architecture. Less than 5% error in was tensile modulus predicted on the basis of the realistic network of composite structure.[3]

The contact and interaction between the polymer matrix and the discontinuous or continuous high stiffness natural fibres and particles play a vital role in determining the properties of the composite. The role of the matrix is chiefly to transmit the stresses to the reinforcement and protect the fibres' surface from mechanical abrasion, retaining the characteristics required for the specific application of the composite.[4,5] The potential use of natural fibres along with polymer matrix composites is mostly in the area of aircraft and automobile interiors, called secondary structures, owing to the composites' high specific stiffness, definite strength, light weight, high corrosion resistance, and fatigue resistance. Composites reinforced with natural fibres have been gaining increased consideration in numerous applications because of their availability, recyclability, biodegradability, and low material cost.[6]

4.2 MODELLING AND SIMULATION OF COMPOSITE STRUCTURES

In order to generate an effective model of progressive failure in thermoplastic reinforced natural fibre composite structures, the damage behaviour should recognize various loading environments. In a compression test, stiffness degradation represents a response to damage and crack spreads in composite laminates. In complex structures, the composite failure mechanisms such as matrix cracking, fibre breaking, and delamination between adjacent plies are used to predict the damage in composite materials.[7] Delamination growth in the matrix is related to various levels of damage such as fracture mechanics, damage mechanics, failure criteria, and damage/plasticity coupling. Continuum damage mechanics (CDM) models deliver a traceable outline for modelling damage initiation and evolution. The Matzenmiller, Lubliner and Taylor (MLT) model of growth for damage variables is sure to follow a Weibull distribution formulation, mainly appropriate for modelling damage propagation phenomena.[8]

Four damage mechanisms for pliable matrix–reinforced fibre composites under compressive load can be distinguished: fibre crushness, resilience of the fibre, plies buckling strength with elastic matrix deformation, failure of the matrix and kink-band. The interfacial debonding is considered as an important degradation mechanism in the composite.[9] The first phase of debonding is initiated by isolated fractures in the weak region.[10] These fractures intensify the stress concentration under the shear effect, causing cyclical rotation of the plies, debonding of the interface, and cracking of the matrix. As the applied load increases, it creates additional cracks between the layers and leads to the ultimate failure of the fibre-reinforced thermoplastic matrix composites.[11] The Cohesive Zone Model was introduced by Dugdale and Yielding and Barenblatt. This model is used for simulating delamination for a wide variety of heterogeneities material at various scales to form the laminate composite.[12]

In the direction of the x axis, the $(1/E_x)$ compliance of an orthotropic unidirectional fibre reinforced lamina considered by its main axes oriented at an angle θ to the coordinate axes would be assessed as[13,14]

$$\frac{1}{E_x} = \frac{\cosh^4\theta}{E_L} + \frac{\sinh^4\theta}{E_T} + \frac{1}{4}\left(\frac{1}{G_{LT}} - \frac{2\,\nu_{LT}}{E_L}\right)\sinh^2 2\theta \qquad (4.1)$$

In Equation 4.1, L and T denote the main material axes (longitudinal and transverse), while E_x represents Young's modulus in the x axis direction. Shear modulus (E_L), Young's moduli (E_T), shear modulus (G_{LT}), and Poisson's ratio (ν_{LT}) are unidirectional fibre-reinforced lamina phenomena. Elastic constants $(E_L, E_T, \nu_{LT},$ and $G_{LT})$ of a unidirectional fibre-reinforced lamina can be found using assumptions and micro mechanics, using known moduli, Poisson's ratios, and volume fractions of the fibre (V_f) and the matrix (V_m) generated as

$$E_L = E_f V_f + E_m V_m \quad V_{LT} = v_f V_f + v_m V_m \qquad (4.2)$$

$$E_T = \frac{E_f E_m}{E_f V_m + E_m V_f} \quad G_{LT} = \frac{G_f G_m}{G_f V_m + G_m V_f} \qquad (4.3)$$

This Halpin-Tsai equation represents the fibre, ribbon or particulate composites and helps in designing composite materials with suitable macroscopic properties.[15] Dissimilar homogeneous isotropic materials and fibre-reinforced composite materials deliver the possibility of tailored mechanical properties by the proper selection of material constituents (matrix and fibre) and their volume amounts. Each lamina is considered as a building block; since several laminae may be stacked one upon another, its ply thickness, orientation of each and individual lamina, and the stacking sequences will decide its mechanical properties. Finally, we may obtain the constitutive calculations for n-layered laminates along with stress-strain relations of an individual fibre-reinforced lamina as per the classical lamination theory.[16,17] Continuous modulus loss owing to hydrolysis reactions of natural fibre–reinforced composites under hygrothermal ageing is considered in order to assess the long-term mechanical behaviour of composites.[18]

4.3   BASICS OF FINITE ELEMENT ANALYSIS

The preprocessor stage includes 1D, 2D, or 3D definitions of the problem; the allocation of the right choice of material and its properties; elements; meshing; limiting appropriate structural boundary settings or thermal settings; and active loads such as electrical loads, thermal loads, magnetic loads, and structural loads. Ansys, Autocad, Catia, Comsol, Creo, Solidworks, Solid Edge, LS-PrePost, and Opti Struct are some of commercial software packages that help to carry out preprocessing modules. In complex geometrical designs using hypermesh software, complex geometries are meshed or interconnected to determine real-time responses to the applied loads (Figure 4.1).

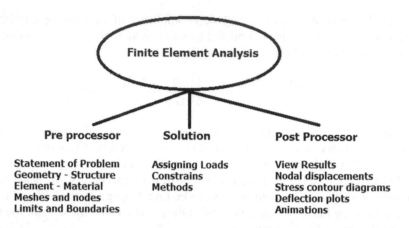

FIGURE 4.1   Overview of finite element analysis process.

In the FEA process, all parameters, such as stiffness, matrix ratio, load, nodal displacements, and element stress and strains, are stored in matrix form. The successful implementation of the meshing process, its global force vectors, element stiffness matrix, and global stiffness matrix are simulated in the back end of the analysis software. The linear equation of an element for a global system is

$$\{f\} = [k]\{u\} \tag{4.4}$$

Here {f} is elemental force vector, [k] is elemental stiffness matrix, and {u} is elemental nodal displacement:

$$\{F\} = [K]\{U\} \tag{4.5}$$

Here {F} is global force vector, [K] is global stiffness matrix, and {U} is global nodal displacement.

If an error appears in FEA, it is resolved using the Gaussian elimination technique back end process. Some of the FEA software are LS-DYNA and ANSYS. The number of elements, nodes, and meshes involved determines its computing time. Higher-order elements would take more computation time compared to linear elements. If any error arises, the entire meshing will need to be cleared and reworked. Depending on the assigning loads, methods and constrains relevant, a new mesh pattern would be considered at this stage.

Postprocessing displays the results of displacements relevant to problems, temperature-based thermal analysis, pressures or velocities-dependent fluid analysis, stresses and strains. It generates the results in terms of graphs, tables, plots, and animations. The responses of a structure under any kind of load, that is, thermal, dynamic, static, impact, fatigue, torque, and so on, can be calculated and analyzed in detail before the actual product is made. LS-PrePost, Opti Struct, ANSYS, and so on are commercially existing software packages for carrying out virtual product simulations.

4.4 DESIGN AND ANALYSIS THROUGH SOFTWARE

Response surface methodology (RSM) is an experiential modelling technique suitable for improving, developing, and optimizing the process, based on the assessment of its relationship to a set of controlled investigational factors and experimental obtained results. The optimization process includes three major steps: the execution of designed statistical experiments, approximating the coefficients of the mathematical model, and forecasting the response and examining the adequacy of the model. RSM method is capable of concurrently determining individual factors and the interactive effects of many factors at the same time.

Table 4.1 shows the thermal properties of reinforced nonwoven composite of jute and hollow conjugated polyester (HCP) as determined by Zakriya et al. in their research work, in which they used an RSM graph to predict the best response and outcome of the interactive effects of associated factors.

4.4.1 EFFECT ON THERMAL CONDUCTIVITY

The thermal conductivity of the nonwoven composite material made up of jute and HCP fibre for all 15 samples, with different fibre proportions and varying thickness, was analyzed and is shown in Figure 4.2A, B and C. It is observed that if the jute fibre content increases, it reduces the λ Lambda value along with an increase of thickness of the composite material. The lowest λ value was obtained in sample S6, at the maximum of 1750 gsm of HCP fibre content with the lowest contribution of

TABLE 4.1
Thermal properties of jute and HCP reinforced nonwoven composite[19]

Sample number	Total weight (g/m²)	Density (kg/m³)	Thermal conductivity (W/mK)	Thermal resistance (m²K/W)	Thermal transmittance (W/m²K)	Thermal diffusivity (m²/s)
S1	3600	720	0.0204	0.245	4.0816	1.8×10^{-8}
S2	3100	689	0.0208	0.2163	4.6232	1.93×10^{-8}
S3	3600	900	0.021	0.1904	5.2521	1.49×10^{-8}
S4	2000	444	0.0317	0.1419	7.0472	4.57×10^{-8}
S5	3700	925	0.0206	0.1941	5.1519	1.42×10^{-8}
S6	3200	933	0.0197	0.2284	4.3782	1.35×10^{-8}
S7	4200	711	0.0231	0.1948	5.1334	2.08×10^{-8}
S8	2600	650	0.0228	0.1754	5.7012	2.24×10^{-8}
S9	3000	667	0.0203	0.2216	4.5126	1.95×10^{-8}
S10	3700	740	0.0205	0.2439	4.1	1.77×10^{-8}
S11	2600	520	0.0224	0.2232	4.4802	2.76×10^{-8}
S12	2500	625	0.022	0.1818	5.5005	2.25×10^{-8}
S13	3100	500	0.0218	0.2293	4.361	2.79×10^{-8}
S14	2500	689	0.0204	0.2205	4.5351	1.89×10^{-8}
S15	3100	689	0.0207	0.2173	4.6019	1.92×10^{-8}

C : Thickness 4.0 mm

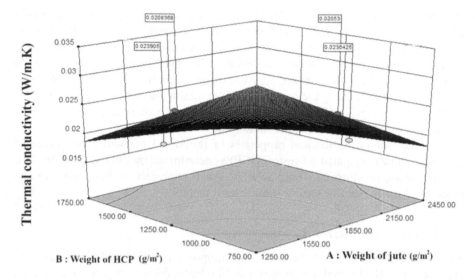

FIGURE 4.2A Thermal conductivity of composite.

C : Thickness 5.0 mm

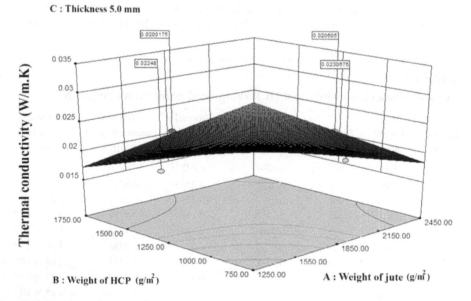

FIGURE 4.2B Thermal conductivity of composite.

1250 gsm of jute fibre. S9 gives $\lambda = 0.0203$ W/mK, and when it is compared with the mean value of 0.02188 W/mK it reduces by 7.2%.

Conversely, a maximum jute fibre content of 1850 and 750 gsm of HCP fibre content in sample S8 with 4 mm thickness and in S11 with 5 mm thickness was considered for evaluation.[19]

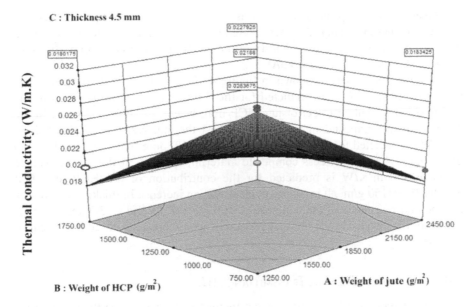

FIGURE 4.2C   Thermal conductivity of composite.

S11 comparatively found 0.0004 W/mK lower differences with greater thickness, which plays a vital role in thermal conductivity qualities. The bulk density of the composite material is directly proportional to its thermal conductivity. Here, the high gsm of Jute/HCP fibre with optimum thickness determines the bulk density of the composites.[19]

From the response surface method (RSM) analysis, the lowest thermal conductivity value 0.0199 W/mK is predicted for the 4 mm thick composite material. Its optimized contribution effect of jute/HCP fibre weight is 1865/1454 g/m^2. In the 4.5 mm thick composite material, the lowest thermal conductivity value 0.0197 W/mK is predicted for the contribution of jute/HCP fibre at a weight of 1934/1363 g/m^2. In the 5 mm thick composite material, the lowest thermal conductivity value of 0.0193 W/mK is predicted for the contribution of jute/HCP fibre that weighs 2097/1260 g/m^2. It is clearly evident that an increase in thickness reduces the thermal conductivity of the composites. In Figure 4.2C, 3D analysis replicates that the curves on the thermal conductivity axis decrease with increments of thickness.[19]

4.4.2   EFFECT ON THERMAL RESISTANCE (R)

An increase in thermal resistance along with an intensification fibre content and a high thickness of 5 mm provides the maximum thermal resistance value, depicted in Figure 4.3 C. Sample S1 possesses the highest R value at 0.245 m^2K/W and sample S4 had the least thermal resistance with a value of 0.1419 m^2K/W. The mean thermal resistance of the composite is 0.20826 m^2K/W; when the density of the composite material decreases the bulkiness of the composite increases relatively, its thickness also increases. High thickness of the composite material improves the thermal resistance of the composite and it acts as a good insulator. Thermal conductance distance increased through thickness increment and it is a reason for reducing the thermal conductivity. The bulk density of the material is indirectly proportional to its thermal resistance[19] (Figure 4.3A).

From the RSM analysis shown in Figure 4.3C, the highest thermal resistance value of 0.2479 m^2K/W is predicted for the 5 mm thick composite material. Its optimized contribution effect of jute/HCP fibre weight is 1786/1552 g/m^2. In the 4.5 mm thick composite material, the highest thermal resistance value 0.2230 m^2K/W is predicted for the contribution of jute/HCP fibre that weighs 1909/1430 g/m^2. In the 4 mm thickness composite material, the highest thermal resistance value 0.2000 m^2K/W is predicted for the contribution of jute/HCP fibre that weighs 2183/1150 g/m^2. It is clearly evident that an increase in thickness increases the thermal resistance of the composites. In Figure 4.3B and C, 3D analysis replicates of the curves on the thermal resistance axis increase with an increment of thickness.[19]

4.4.3 EFFECT ON THERMAL TRANSMITTANCE (U)

Jute fibre content and HCP fibre content increases along with the thickness factor which reduces the thermal transmittance value of the nonwoven composite material; this is clearly shown in Figure 4.4 A, B & C. For a material with good insulation properties, the rate of transfer of heat through the composite material from its differences in temperature across the composite material should be low. Sample S1 possess the lowest thermal transmittance value at 4.0816 W/m^2K, with the next lowest being that of sample S10 with a U value of 4.1 W/m^2K.[19]

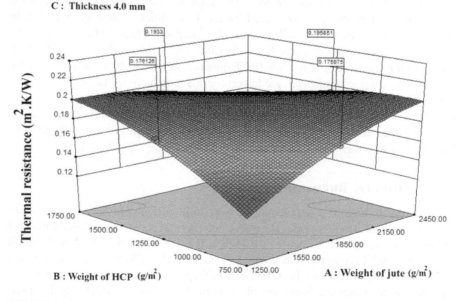

FIGURE 4.3A Thermal resistance of composite.

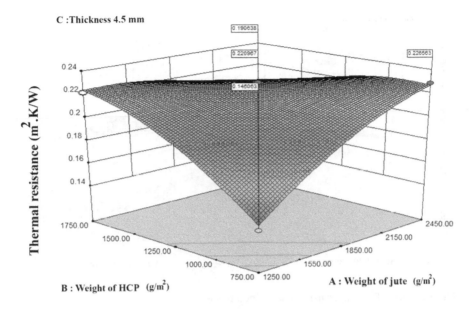

FIGURE 4.3B   Thermal resistance of composite.

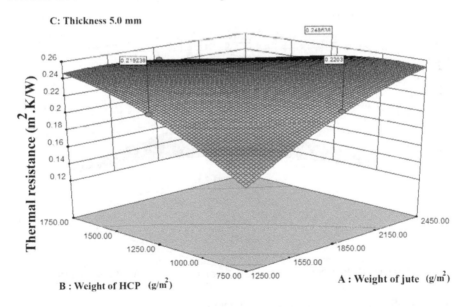

FIGURE 4.3C   Thermal resistance of composite.

The thickness of the composites plays a vital role in determining the U value. Reducing the density of the material by adjusting the needling density on the nonwoven manufacturing process and compression pressure regulation can determine the thickness of the composite material.[19]

From the RSM shown in Figure 4.4C, the lowest thermal transmittance value 3.8695 W/m²K is predicted for the 5 mm thick composite material. Its optimized

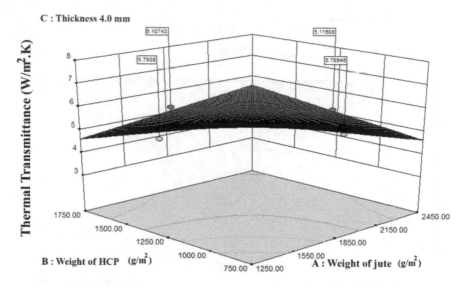

FIGURE 4.4A   Thermal transmittance of composite.

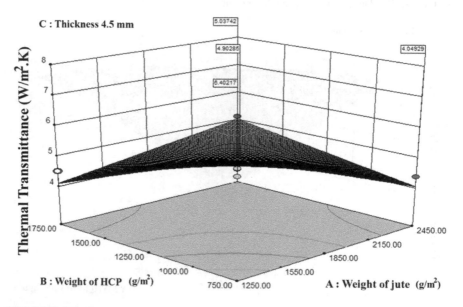

FIGURE 4.4B   Thermal transmittance of composite.

contribution effect of jute/HCP fibre weight is 1815/1480 g/m^2. In the 4.5 mm thick composite material, the lowest thermal transmittance value 4.4595 W/m^{2}K is predicted for the contribution of jute/HCP fibre that weighs 1892/1425 g/m^2. In the 4 mm thickness composite material, the lowest thermal transmittance value of 4.4596 W/m^{2}K is predicted for the contribution of jute/HCP fibre that weighs 1963/1370 g/m^2.

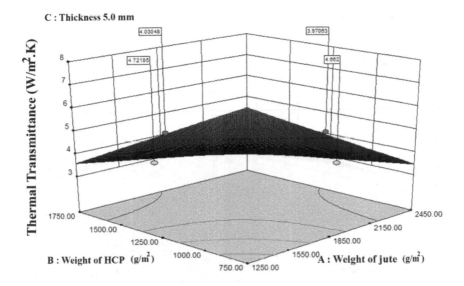

FIGURE 4.4C Thermal transmittance of composite.

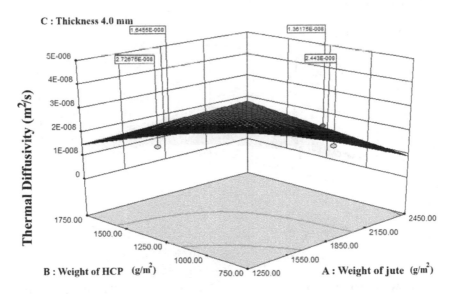

FIGURE 4.5A Thermal diffusivity of composite.

It is evident that an increase in thickness reduces the thermal transmittance of the composites. In Figure 4.4B and C, 3D analysis replicates of the curves on thermal transmittance axis decrease with increments of thickness.

4.4.4 Effect on Thermal Diffusivity (α)

Low thermal diffusivity slows down the flow of heat inside the nonwoven composite and is related to the bulk density of the composite material. Figure 4.5A, B and C

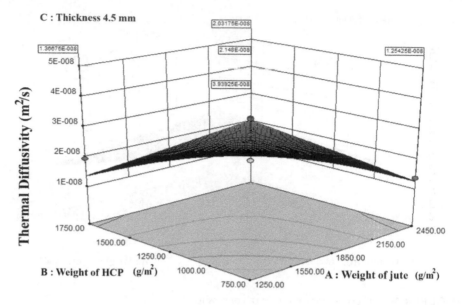

FIGURE 4.5B   Thermal diffusivity of composite.

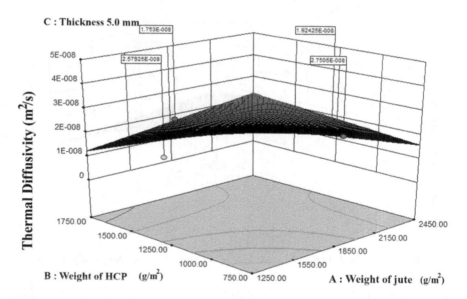

FIGURE 4.5C   Thermal diffusivity of composite

show that there are no more changes directly related to the thickness of the composite material. Sample S6 acquired the lowest thermal diffusivity value of 1.35×10^{-8} m^2/s because it possesses 933 kg/m^3 density. Contrarily, thickness reduction with respect to higher thermal transmittance improves the bulk density of the composite. The improvement of bulk density reduces the thermal diffusivity of the composite material. Air gaps present in between layers of the sandwich structure and the air pockets within the HCP fibres reduce thermal conduction in a bulged composite material.[19]

From the RSM analysis shown in Figure 4.5C, the lowest thermal diffusivity value of 1.48×10^{-8} m²/s is predicted for a 4 mm thick composite material. Its optimized contribution effect of jute/HCP fibre weight is 2301/1141 g/m². In the 4.5 mm thick composite material, the lowest thermal diffusivity value 1.48×10^{-8} m²/s is predicted for the contribution of jute/HCP fibre that weighs 2402/945 g/m². In the 5 mm thickness composite material, the lowest diffusivity value of 1.45×10^{-8} m²/s is predicted for the contribution of jute/HCP fibre that weighs 1419/1719 g/m². It is evident that an increase in thickness reduces the thermal diffusivity of the composites. In Figure 4.5C, 3D analysis replicates of the curves on thermal diffusivity axis decrease with increments of thickness. In a fluctuating thermal situation, the thermal diffusivity of a composite material determines the dissemination depth and the speed of temperature adaptation of the nonwoven composite.[19]

4.5 CONCLUSION

The autoclave process method is extensively used in the manufacture of thermosetting matrices with natural fibre–reinforced composite materials. Nonuniformity of temperature transfer or spreads in autoclave moulds affect synchronous curing of the composite and reduce the mechanical properties of composites; it can be avoided by the design optimization of autoclave moulds. In order to envisage the mechanical properties, it is simpler to design and description on the basis of models than to carry out experiments.[20] The tool of FEA is supported to simulate each phase of the composite separately, including fibres types and its structures, interface arrangement and its matrix interphases.[21] Fibre–matrix debonding, the effects of matrix cracking, and changes in the microstructure of natural fibres on mechanical performance of the composite are taken into account. At the same time, the existence of competition mechanisms in the evolution of mechanical properties of natural fibres can be determined through modelling and simulations.

REFERENCES

1. Ameen, M. *Boundary Element Analysis: Theory and Programming.* Boca Raton, FL: CRC Press, 2001.
2. Harper, L.T, Qian, C.C., Luchoo, R. and Warrior, N.A. 3D geometric modelling of discontinuous fibre composites using a force-directed algorithm. *Journal of Composite Materials* 51(17), 2389–2406, 2017.
3. Wang, Q, Wang, L., Zhu, W., Xu, Q. and Ke, Y. Design optimization of molds for autoclave process of composite manufacturing. *Journal of Reinforced Plastics and Composites* 36(21), 1564–1576, 2017.
4. Nicolais, L., Gloria, A. and Ambrosio, L. The mechanics of biocomposites. In Ambrosio, L. (ed.). *Biomedical Composites.* Cambridge, UK: Woodhead Publishing Limited, 2010, pp. 411–440.
5. Gloria, A., De Santis, R. and Ambrosio, L. Polymer-based composite scaffolds for tissue engineering. *Journal of Applied Biomaterials & Biomechanics* 8, 57–67, 2010.
6. Naveen, J., Jawaid, M., Vasanthanathan, A. and Chandrasekar, M. Finite element analysis of natural fibre-reinforced polymer composites. In Jawaid, Mohammad, Thariq,

Mohamed and Saba, Naheed (eds.). *Modelling of Damage Processes in Biocomposites, Fibre-Reinforced Composites and Hybrid Composites.* UK: Woodhead Publishing, 2019.

7. Aziz, V.D. and Tsai, S.W. Anisotropic strength of components. *Experimental Mechanics* 5, 286–298, 1965.

8. Mokhtari, A., Ould Ouali, M. and Tala-Ighil, N. Damage modelling in thermoplastic composites reinforced with natural fibres under compressive loading. *International Journal of Damage Mechanics* 24(8), 1239–1260, 2015.

9. Jelf, P.M. and Fieck, N.A. Compression failure mechanisms in unidirectional composites. *Journal of Composite Materials* 26(18), 2706–2721, 1992.

10. Maimi, P., Camanho, P.P., Mayugo, J.A. and Dávila, C.G. A continuum damage model for composite laminates: Part I -constitutive model. *Mechanics of Materials* 39, 897–908, 2007.

11. Maimi, P., Camanho, P.P., Mayugo, J.A. and Dávila, C.G. A continuum damage model for composite laminates: Part II – computational implementation and validation. *Mechanics of Materials* 39(10), 909–919, 2007.

12. Dugdale, D.S. and Yielding, J. In steel sheets containing slits. *Journal of the Mechanics and Physics of Solids* 8, 100–104, 1960.

13. Hull, D. *An Introduction to Composite Materials.* USA: Cambridge University Press, 1981.

14. Ambrosio, L., Gloria, A. and Causa, F. Composite materials for replacement of ligaments and tendons. In: Ambrosio, L. (ed.) *Biomedical Composites.* Cambridge, UK: Woodhead Publishing Limited, 2010, pp. 234–254.

15. Gloria, A., Ronca, D., Russo, T., D'Amora, U., Chierchia, M., De Santis, R., Nicolais, L. and Ambrosio, L. Technical features and criteria in designing fibre-reinforced composite materials: From the aerospace and aeronautical field to biomedical applications. *Journal of Applied Biomaterials & Biomechanics* 9(2), 151–163, 2011. doi: 10.5301/JABB.2011.8569

16. Jones, R.M. *Mechanics of Composite Materials* (2nd ed.). London, UK: Taylor & Francis, 1999.

17. Nicolais, L., Gloria, A. and Ambrosio, L. The mechanics of biocomposites. In Ambrosio, L. (ed.). *Biomedical Composites.* Cambridge, UK: Woodhead Publishing Limited, 2010, pp. 411–440.

18. Tian, F., Zhong, Z. and Pan, Y. Modeling of natural fibre reinforced composites under hygrothermal ageing. *Composite Structures* 200, 144–152, 2018. doi: 10.1016/j.compstruct.2018.05.083

19. Zakriya, G.M., Ramakrishnan, G., Palani Rajan, T. and Abinaya, D. Study of thermal properties of jute and hollow conjugated polyester fibre reinforced non-woven composite. *Journal of Industrial Textiles* 46(6), 1393–1411, 2017. doi: 10.1177/1528083715624258

20. Cao, Y., Wang, W. and Wang, Q. Application of mechanical model for natural fibre reinforced polymer composites. *Materials Research Innovations* 18, S2–357, 2014.

21. Chegdani, F., El Mansori, M., Bukkapatnam, S.T. and Reddy, J.N. Micro mechanical modeling of the machining behavior of natural fibre-reinforced polymer composites. *The International Journal of Advanced Manufacturing Technology*, 105, 1549–1561, 2019. doi: 10.1007/s00170-019-04271-3

5 Process and Production Techniques of Composites

5.1 INTRODUCTION

A composite is a blend of two main elements, one called the *matrix* and the other *reinforcement*. Composites can be classified into three categories: metal matrix composites (MMC), ceramic matrix composites (CMC), and polymer matrix composites (PMC). In PMC, fibres are mingled with resin or plastic to become fibre reinforced plastic (FRP).[1] Micro cracks in the interface between fibres and matrix lead to a substantial loss of composite stiffness and strength. Multi directional preforms made up of reinforced fibres yield cost effectiveness, damage tolerance and better structural integrity, wide structural design possibility, superior out-of-plane belongings and especially able to adopt printing spatially oriented fibres on composite manufacturing.[2–5]

In the past few years, numerous research groups around the world have fabricated effective materials using biodegradable fibres and polymers. These kinds of biocomposites are called *green* composites. Many countries have established sets of policies to encourage businesses to produce biocomposites that can help save energy and materials and sustain and maintain the environment.[6] Tax incentives are an alternative way to encourage different manufacturing firms to design and make successful products with 'green' components. In recent times, bamboo fibre has attracted more consideration than other natural fibres. Its fast growth rate, low cost, low density, and high mechanical strength are its advantages. The most outstanding virtue of bamboo is its ability to generate oxygen and absorb carbon dioxide at a rate that is approximately three times higher than other plants, which helps to minimize the greenhouse effect to a certain extent.[7]

The nature of natural fibre–reinforced polymer composites nature is determined by the selection criteria used in the manufacturing process. It depends upon the complexity of the final products, dimensional tolerance, production time, and volume of production.[7] Regardless of the many factors at play, cost competitiveness and the renewability of natural fibres are encouraging business sectors to explore the opportunities offered by natural fibre composites. The appropriate selection of a composite fabrication process will help to make the most of structural attributes such as desired component, size, and shape in manufacturing industries in the future.[8–10] A range of natural fibres and matrices are indicated in Table 5.1.

Small quantities of additives can significantly develop the physical, chemical, or mechanical properties of the composite. Improvements in the interfacial bond between fibre and matrix heighten the performance of the composites.[12,13] Various

TABLE 5.1
Different kinds of polymer matrices[11]

Polymer	Matrices
Ketone	Thermoplastic matrices
Polyethylene	
Polystyrene	
Nylons	
Polycarbonate	
Polyacetals	
Polyamide-imide	
Polyether-ether	
Epoxies	Thermosetting
Phenol	matrices
Polyesters	
formaldehyde	
Butyl rubber	Rubber matrices
Butadiene rubber natural rubber	
Styrene butadiene rubber	
Shellac	Biodegradable matrices
Starch	
Proteins	
Polyhydroxyalkanoates	

treatments, such as chemical reactions, surface modifications, pre-impregnation, and plasma have been used to improve interfacial shear strength of composites.[14] Chemical treatments such as chemical grafting, delignification, dewaxing, and acetylation are used to modify the surface characters of the fibres, thereby improving composite performance. Plasma treatment is an environmentally friendly technology that helps to alter the surface properties of composites without altering their bulk properties.[15] The size and dimension of the composites is an initial factor to be considered when selecting a suitable production process. Injection and compression mouldings are preferred for their simplicity and fast processing cycle for small and medium-size components. Autoclave, pultrusion, and open moulding processes are employed for manufacturing large structures.[16]

5.2   FIBRE REINFORCEMENTS

Fibre reinforcement can be done by several methods; some of the commercial methods are as follows:

5.2.1   SPRAY METHOD

A spray gun is used to spray the surface of the mould with a mixture of chopped fibres, resin, and catalyst. This open-mould process is a semi-automatic process. It is shown in

Figure 5.1. Spray rollers are used to squeeze out excess resin and remove entrapped air inside the fibres. The presence of resin-rich heavy volatile organic compounds creates air pollution. This spray gun method can also be used in a closed moulding technique, with the help of closed mould plates both sides of the composite surface get well finished. This method is used to make medium to large complex forms at low cost.[17,18]

5.2.2 LAMINATION

It can be done as a hand lay-up process or be automated while using sheet-moulding compounds. The thickness of the final laminate composite determines the number of required layers. It can be cured at high temperatures or at room temperature depending on the resin system utilized. It is shown in Figure 5.2. To achieve better compaction, a pressure plate or vacuum bagging is used. The result is a low fibre volume fraction attained at a lower cost.[18]

5.2.3 FILAMENT WINDING

Filament winding encompasses winding and laying processes: cloth winding or wrapping and tape laying or wrapping. This is clearly depicted in Figure 5.3. This method is

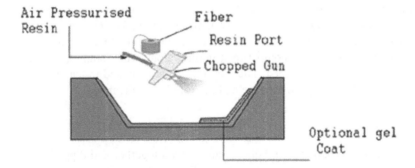

FIGURE 5.1   Spray coating process.[17]

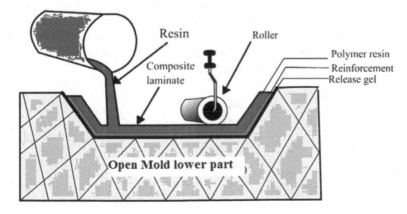

FIGURE 5.2   Hand lay-up process.[17]

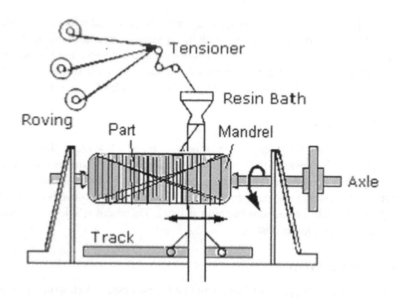

FIGURE 5.3   Filament winding process.[17]

used to fabricate axially symmetric fibre-reinforced structures. The process involves passing the fibre through a liquid resin or matrix solution; the pre-impregnated fibre is then wound onto a proper mandrel. The fibres or tape or cloth are wrapped at differ-ent orientations. The entire composite structure, along with mandrel, is cured. After curing, the mandrel is detached. The winding angle is related to the combination of translation of the transverse carriage and the rotation of the mandrel.[19,20]

5.2.4   PULTRUSION

Continuous fibres pulled from bobbins are guided together and dipped into a resin tank, before being passed through a heater. The material exits through a heated die, which gives the desired shape before final curing. Pulling mechanism in the system able to provide cutting edge desired shape final products. This kind of process is used to produce high strength continuous cross sections for automated mass produc-tion systems.[21,22] Due to the resulting good surface quality and full automation, this technique is suited to the automotive and aerospace industries.[23] The resin transfer moulding process is depicted in Figure 5.4.

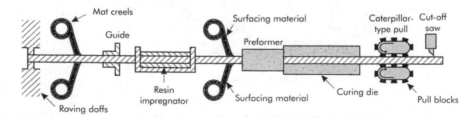

FIGURE 5.4   Resin transfer moulding process.[17]

5.2.5 COMPRESSION MOULDING

The core factors in the compression moulding process are time, pressure, and temperature. The factors are decided based on final thickness, size and shape of composite material and its curing cycle requirement. There are various methods of compression moulding, including sheet moulding compound (SMC), thick moulding compound (TMC), bulk moulding compound (BMC), and wet layup moulding compound (WMC). In wet moulding, the top die is movable and the bottom is stationary. Pressure is applied by a hydraulic press. It is clearly shown in Figure 5.5. The temperature level is controlled and regulated via an electrical circuit. Optimum pressure and temperature control between the top die and bottom die determines the quality of composite. Variations in temperature, pressure, and time help to achieve the desired property for the composite. This method is suitable for producing excellent similar repeatable quality composites with less trimming cost.[24]

5.2.6 VACUUM BAGGING AND VACUUM INFUSION

Vacuum bagging adopts a wet hand lay-up process covered with plastic film and vacuum is drawn out from the sealed edges. Due to vacuum pressure, excess resin and voids are squeezed out of the composite structure.[25]

Vacuum infusion process is used in dry lay-up process covered with vacuum bagging materials and infused resins in the mould. It is generally used to make highly finished large structures and is also known as vacuum assisted resin transfer moulding (VARTM). It is shown in Figure 5.6. The process is highly suitable for the production of aviation components, boat components, and large auto body panels.[25]

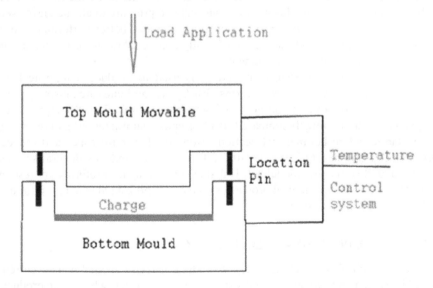

FIGURE 5.5   Compression moulding process.[17]

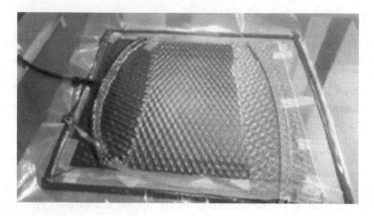

FIGURE 5.6 VARTM process.[17]

5.3 FORM REINFORCEMENTS

Prepreg is a reinforcement pre-soaked with a resin and slightly dried to escalate the viscosity. Wet prepreg layers are placed on a specified shape mould by hand. Uniform composites are produced by removing trapped air during the process. Pressure is typically generated through hand rolling or vacuum bagging. Fibres are typically prepared in the form of unidirectional tape/woven fabric soaked with cured resin. This method is much more precise than the wet lay-up method processed through autoclaving and vacuum bagging processes.[26]

Bridging is formed if some parts of the composite areas are not appropriately compressed. Particularly in deep and narrow channels, vacuum bagging creates bridges across the channels. By using an autoclave, it is possible to provide pressure values higher than in a vacuum. Metal and composite moulds used in autoclaves withstand the forces during the process. The heat and pressure in an autoclave eventually create a vacuum. Curing in an autoclave provides better performance than non-pressure techniques. The autoclave is mostly used to fabricate very complex and high-performance aerospace components.[26]

Vacuum-assisted resin transfer moulding is used to produce composite laminates reinforced with woven silk preforms. Silk/epoxy laminates are anisotropic and equivalent to glass/epoxy composite. Silk composites exhibited a 23% improvement in specific flexural strength compared to glass/epoxy laminates. Moisture present in silk fabric before laminate fabrication results in slower fill times and reduced mechanical properties. On average, 10% fabric moisture reduces flexural strength by 25% and 20% specific modulus.[27] Unbalanced woven silk preforms are responsible for anisotropic mechanical properties, and developing a balanced preform may address the anisotropic issue.[27]

5.4 NEEDLE PUNCHING TECHNOLOGY

In their research, Zakriya et al. made 04 kinds of needle-punched nonwoven reinforced jute and hollow conjugated polyester (HCP) composites, which were produced using the following methods:[28]

5.4.1 Structure I (Sandwich Type)

Jute fibre was opened out using a pilot Trytex card machine at the doffer speed of 4 rad/s. Jute webs were produced in the form of sheets almost 0.4 m in width. Similarly, HCP individualized stapled fibre web sheets were produced separately. Randomized fibre arrangement of the web was arranged into following HCP–Jute–HCP–Jute–HCP sandwich layers. As per the desired level of (g/m^2) gsm requirement several layers of web were arranged one upon another.

A Dilo, a German-made needle-punching machine with a working width of 8,000–10,000 needles/m and a downstroke board, was used to prepare the nonwoven material. Needling density was maintained at 300 punches/cm^2 at a speed of 257 cycles/min (30 strokes) and a constant penetration depth of 11 mm set for the barbed needle, which had a dimension of 15 × 18×36 × 3.5 R/SP (rounded point/slightly rounded point).

5.4.2 Structure II (Blended Structure)

In a RE 5 machine, jute and hollow conjugated polyester fibres were opened out into small tufts and fed into a card feeder. The opener roller speed and feed percentage were set at 350 rpm and 24%. Material received from the fibre opener was passed through a chute feed system, and the blended fibre material was opened up and web formed in a card feeder CFL 7 machine. Here the feed speed was limited to 20%. The blended structure manufacturing process is shown in Figure 5.7.

The web was then passed through a roller carding CAL 7 machine. The batt produced from the card feeder was opened and carded, which produced a thin sheet of carded web. The inlet speed in the carding machine was set at 0.59 m/min, the cylinder speed at 125 m/min, and the delivery speed at 17 m/min. The webs produced in the lap layer F7/6 machine were cross laid, forming batt by using a draft transfer position at 2% and layering draft at 0.5% and adjusting the layering factor from 24 to 34 according to required g/m^2 of the batt material. Finally, the batt was sent to the

| (a) Blended Jute/HCP batt before needling | (b) Blended Jute/HCP batt after needling |

FIGURE 5.7 Manufacturing process of blended nonwoven structure.

needle punching machine model DI-Loom OUG-II 6, which has a working width of 600 mm with 6000–20,000 needles/m. A maximum of 1200 strokes/min at 60 mm penetration depth was maintained. The machine has two needle boards at down-stroke and two needle boards at upstroke, placed opposite each other. In total, it had four boards for double-sided needling, and the boards had a barbed needle dimension of 15 × 18 × 36 × 3.5 R/SP.

5.4.3 STRUCTURE III (BLENDED FIBRE WITH 5% OF LOW MELT POLYESTER)

The aforementioned procedure is followed to make blended fibre nonwoven samples as shown in Figure 5.6. Instead of producing a single structure, multiple layers of batt are formed. In between the layers of batt, 5% of total weight of low-melt polyester fibre is added. The batt is then brought to the needle punching process.

5.4.4 STRUCTURE IV (MULTIPLE LAYER NONWOVENS)

Instead of having low-melt polyester fibre added to it, the blended fibre (50/50) is processed at a fixed needling density. Multiple layers of nonwovens are prepared and arranged one on top of the other for the stitching process, until the fixed weight of the material is reached. A schematic view of this kind of composite preparation is shown in Figure 5.8.

A compression moulding technique is adopted to make a nonwoven composite. A hot pressing temperature of 160°C is maintained to initiate the thermo-bonding process of the webs, and a pressing time of up to 30 min is allowed. HCP fibre is a bi-component fibre which consists of a sheath and a core. The sheath fibre component starts melting at 110°C. The pressure level is maintained at 0.6 ± 0.02 MPa, and after the fabrication process, the pressure is relaxed so that the sample can be collected from the compression moulding device.

The nonwoven sandwich structure produced by the compression moulding process is itself bonded together and transformed into sandwich composites. The blended nonwoven material, with its sheath of bi-component fibre, is bonded with jute fibre to form a homogeneous matrix composite. The addition of low melt polyester fibre content to the blended nonwoven materials creates a spontaneous bond with each layer, improving the strength of composite and making it rigid. Multiple-layer

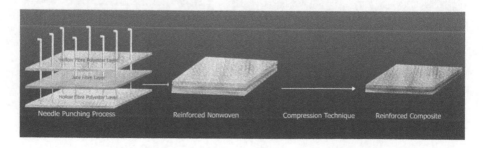

FIGURE 5.8   Schematic view of composite manufacturing process.[29]

nonwovens without low-melt polyester fibre are stitched with 60^s Ne two-ply nylon sewing thread to form strong flexible composites.[28]

5.5 MATERIAL SELECTION IN DESIGN

The superiority and strength of natural fibre–reinforced composites are dependent upon the fibres' maturity and size and the technique used to process them. The extraction of natural fibres from animal or plants through processes such as dew-retting, scutching, ripping, cleaning, blending or mixing, carding, combing, draw-frame doubling, and spinning determines the inherent characteristics of the composite material. Fibres with high cellulose content and low micro-fibrillar angles create the best results in composite preparation.[30] Fibres like flax, pinus, hibiscus, jute, sisal, and sabdariffa have proven to be good reinforcement fibres in thermoset and thermoplastic matrices and also exhibit good thermal insulation characteristics and mechanical attributes.[31-33]

The interlacing pattern of the weave structure and how tightly the fibres are attached together in the woven structure are important criteria in improving the strength of composites.[34] Researchers have found that the matt type of woven composite shows better mechanical properties compared to plain and twill weave–based composites.[35] Manmade antimicrobial agents such as cadmium, copper, silver, and mercury hold a large portion of market share in antimicrobial textiles. These contain heavy metal complex compounds, which can be added to or processed together with textile materials. Produced IN forms such as fibre hunch, carded web, and yarn or fabric, or preformed into durable composite structures, these textiles will act as inhibitors for the growth of microorganisms in composite structures.[36,37]

Over the past decade, interest has grown in blending PLA fibre with other polymers such as PVP (polyvinyl pyrolidone), PCL (poly e-caprolactone), starch, chitosan, PHB (poly 3-hydroxybutyrate), PVB (polyvinyl butyral), PBS (polybutylene succinate), rice starch (RS), Arabic gum (AG), PEG (polyethylene glycol), PBSA (polybutylene succinate adipate), and PBAT (polybutylenes adipate-co-terephthalate); these polymers are widely considered to produce biodegradable medical-aided composite structures.[38-41] Biodegradable polymers are defined as ones in which degradation occurs naturally by means of micro-organisms like algae, bacteria, and fungi. Some examples of such polymers are polyhydroxyl alkanoate, polylactide, polycaprolactone, cellulose acetate, cellulose propionate, and cellulose butyrate.[14,42]

Chicken feather fibres have the potential to produce good acoustic and thermal insulated composite materials. Due to its low-density fibre and biodegradability, this fibre may be combined with appropriate biopolymer matrices to make biocomposites with interesting potential for industrial applications.[43] Nowadays, biochar, calcium carbonate with diammonium phosphate, wool, and so on are blended with other natural fibres to made composite materials flame retardant. A high-density material requires a higher level of oxygen for its combustion. Chemical surface finishes can also be applied to composite surfaces to make them flame retardant.[44]

5.6 CONCLUSION

Composite materials are chosen for their superiority in terms of strength and stiffness in comparison to conventional metallic alloys. An added advantage is their low coefficient of thermal expansion and enhanced directional properties.[45] Each individual natural fibre has very specific properties and may replace or reduce utilization of synthetic fibres according to requirements. Natural fibre–reinforced thermoplastic composites are superior to conventional materials due to their fast production cycle, ease of processing, and low tooling cost. They make a very suitable material for the requirements of the automobile and electrical industries. The enriched material performance depends on the strength of the interfacial bond between the fibres and the matrix. As per requirements, the polymer in layer contact with the fibre surface can be changed by changing the bulk density of the composite material. Fibre–polymer interactions due to immobilization of the matrix chains, electrostatic forces, presence of chemical bonds, internal stresses, voids or micro-cracks in the interlayer, gaps between lamina and lot more options are available to a researcher to make custom tailoring in order to meet the desired property on the composites by adopting different kinds of manufacturing technologies.[46] Preforms made using knitting, weaving, and nonwoven technology are considered to provide an even surface for the distribution of the matrix in composite manufacturing to address the anisotropic issues.[47]

REFERENCES

1. Hollaway, L.C. A review of the present and future utilization of FRP composites in the civil infrastructure with reference to their important in-service properties. *Construction and Building Materials* 24, 2419–2445, 2010.
2. Van der Zwaag, S., Grande, A., Post W., Garcia, S.J. and Bor, T.C. Review of current strategies to induce self-healing behaviour in fibre reinforced polymer based composites. *Materials Science and Technology* 30, 1633–1641, 2014.
3. Norris, C., Bond, I. and Trask, R. The role of embedded bio inspired vasculature on damage formation in self-healing carbon fibre reinforced composites. *Composites Part A: Applied Science and Manufacturing* 42, 639–648, 2011.
4. Hamilton, A.R., Sottos, N.R. and White, S.R. Self-healing of internal damage in synthetic vascular materials. *Advanced Materials* 22, 5159–5163, 2010.
5. Quan, Z., Wu, A. and Keefe, M. Additive manufacturing of multi-directional preforms for composites: Opportunities and challenges. *Material Today* 18, 503–512, 2015.
6. Vaisanen, T., Haapala, A., Lappalainen, R. and Tomppo, L. Utilization of agricultural and forest industry waste and residues in natural fibre–polymer composites: A review. *Waste Management* 54, 62–73, 2016.
7. Laua, K.T., Hung, P.Y., Zhu, M.H. and Hui, D. Properties of natural fibre composites for structural engineering applications. *Composites Part B* 136, 222–233, 2018.
8. Azaman, M.D., Sapuan, S.M., Sulaiman, S., Zainudin, E.S. and Abdan, K. An investigation of the process-ability of natural fibre reinforced polymer composites on shallow and flat thin-wall parts by injection moulding process. *Materials and Design* 50, 451–456, 2013.
9. George, M. and Bressler, D.C. Comparative evaluation of the environmental impact of chemical methods used to enhance natural fibres for composite applications and glass fibre based composites. *Journal of Cleaner Production* 149, 491–501, 2017.

10. Rahman, M.A., Parvin, F., Hasan, M. and Hoque, M.E. Introduction to manufacturing of natural fibre-reinforced polymer composites. In Salit, M., Jawaid, M., Yusoff, N. and Hoque, M. (eds.). *Manufacturing of Natural Fibre Reinforced Polymer Composites.* Cham: Springer, 2015, pp. 17–43.

11. Kozlowski, R. and Machiewacz Talarczyk, M. Composites reinforced with ligno-cellulosic fibres. In Nor, M.Y.M. (ed.). *Proceedings: Eight Pacific Rim Bio-Based Composite Symposium, Kuala Lumpur, Malaysia,* 20–23 November. Malaysia: Institut Penyelidikan Perhutanan, 2006, pp. 12–28.

12. Ho, M.P., Wang, H., Lee, J.H., Ho, C.K., Lau, K.T., Leng, J. and Hui, D. Critical factors on manufacturing processes of natural fibre composites. *Composites Part B* 43, 3549–3562, 2012.

13. Matuana, L.M. and Balatinecz, J.J. Surface characterization of esterified cellulosic fibres by XPS and FTIR spectroscopy. *Wood Science and Technology* 35, 191–201, 2001.

14. Saheb, D.N. and Jog, J.P. Natural fibre polymer composites: A review. *Advances in Polymer Technology* 18(4), 351–363, 1999.

15. Keener, T.J. and Stuart, R.K. Maleated coupling agents for natural fibre composites. *Composites Part A* 35, 357–362, 2004.

16. Rajmohan, T., Vinayagamoorthy, R. and Mohan, K. Review on effect machining parameters on performance of natural fibre–reinforced composites. *Journal of Thermoplastic Composite Materials* 864–874, 2018. doi: 10.1177/0892705718796541

17. Alpa Tapan Bhatt, Gohil Piyush P. and Chaudhary, Vijaykumar 2018 Primary manufacturing processes for fiber reinforced composites: History, development & future research trends, *IOP Conference Series: Materials Science and Engineering* 330, 012107. https://iopscience.iop.org/article/10.1088/1757-899X/330/1/012107

18. Huntsman International LLC. Wet Lay-Up Moulding Processes, Retrieved February 2020 from https://www.huntsman.com/advanced_materials/a/Our%20Technologies/Ready%20to%20Use%20Formulated%20Systems/Composites/Wet%20Lay-Up, 2001–2020.

19. Yang, B.J., Ha, S.K., Pyo, S.H. and Lee, H.K. Mechanical characteristics and strengthening effectiveness of random-chopped FRP composites containing air voids. *Composites: Part* 62, 159–166, 2014.

20. Hernández-Moreno, H., Douchin, B., Collombet, F., Choqueuse, D. and Davies, P. Influence of winding pattern on the mechanical behavior of filament. *Composites Science and Technology* 68(3–4), 1015–1024, 2008.

21. Adrian, P.P., Gheorghe, B.M. Manufacturing process and applications of composite materials. Annals of the Oradea University. *Fascicle of Management and Technological Engineering* IX(XIX), NR2, 2010.

22. Rufe, P.D. *Fundamentals of Manufacturing* (2nd ed.). Dearborn, MI: Society of Manufacturing Engineer Editor, 2002.

23. Dexter, H.B. Development of textile reinforced composites for aircraft structures. In *Proceeding of the 4th International Symposium for Textile Composites,* pp. 0-32-1-0-32-8. Kyoto, Japan, 12–14 October, 1998.

24. Composite material and manufacturing by society of manufacturing Engineering, Retrieved January 2020 from https://www.sme.org/technologies/manufacturing-topics/composite-materials-manufacturing/

25. Strong, B.A. Manufacturing methods. In Ploskonka, C.A. (Ed.). *Fundamentals of Composites Manufacturing: Materials, Methods, and Applications.* Dearborn, MI: Society of Manufacturing Engineers, Publications Development Dept., Reference Publications Division, 1989, pp. 107–159.

26. Gloria, A., Ronca, D., Russo, T., Amora, U., Chierchia, M., De Santis, R., Nicolais, L. and Ambrosio, L. Technical features and criteria in designing fibre-reinforced

composite materials: From the aerospace and aeronautical field to biomedical applications. *Journal of Applied Biomaterials and Biomechanics* 9(2), 151–163, 2011. doi: 10.5301/JABB.2011.8569

27. Hamidi, Y.K., Akif Yalcinkaya, M., Guloglu, G.E., Pishvar, M., Amirkhosravi, M. and Cengiz Altan, M. Silk as a natural reinforcement: Processing and properties of silk/epoxy composite laminates. *Materials* 11, 2135, 2018. doi: 10.3390/ma11112135

28. Zakriya, G.M. and Ramakrishnan, G. Insulation and mechanical properties of jute and hollow conjugated polyester reinforced nonwoven composite. *Energy and Buildings* 158, 1544–1552, 2018. doi: 10.1016/j.enbuild.2017.11.010

29. Mohamed Zakriya, G. and Ramakrishnan, G. Jute and hollow conjugated polyester composites for outdoor & indoor insulation applications. *Journal of Natural Fibres* 16(2), 185–198, 2019. doi: 10.1080/15440478.2017.1410515

30. Hosur, M., Maroju, H. and Jeelani, S. Comparison of effects of alkali treatment on flax fibre reinforced polyester and polyester-biopolymer blend resins. *Polymers & Polymer Composites* 23(4), 229–242, 2015.

31. Noorunnisa Khanam, P., Abdul Khalil, H.P.S., Jawaid, M., Reddy, G.R., Narayana, C.S. and Naidu, S.V. Sisal/carbon fibre reinforced hybrid composites: Tensile, flexural and chemical resistance properties. *Journal of Polymers and the Environment* 18, 727–733, 2010.

32. Gujjala, R., Ojha, S., Acharya, S. and Pal, S.K. Mechanical properties of woven jute-glass hybrid-reinforced epoxy composite. *Journal of Composite Materials* 48(28), 3445–355, 2014.

33. Kumar, M.A., Reddy, G.R., Bharathi, Y.S., Naidu S.V. and Naga Prasad Naidu V. Frictional coefficient, hardness, impact strength, and chemical resistance of reinforced sisal-glass fibre epoxy hybrid composites. *Journal of Composite Materials* 44, 3195–3202, 2010.

34. Yahaya, R., Sapuan, S.M., Jawaid, M., Leman, Z. and Zainudin, E.S. Mechanical performance of woven Kenaf-Kevlar hybrid composites. *Journal of Reinforced Plastics and Composites* 33(24), 2242–2254, 2014.

35. Khan, G.M.A., Terano, M., Gafur, M.A. and Alam, M.S. Studies on the mechanical properties of woven jute fabric reinforced poly (l-lactic acid) composites. *Journal of King Saud University-Engineering Sciences* 28(1), 69–74, 2016.

36. Purwar, R. and Joshi, M. Recent developments in antimicrobial finishing of textiles–A review. *AATCC Review* 4, 22–26, 2004.

37. Ramachandran, T., Rajendrakumar, K. and Rajendran, R. Antimicrobial textiles and overview. *Journal of the Institution of Engineers* 84, 42–47, 2004.

38. Broz, M.E., VanderHart, D.L. and Washburn, N.R. Structure and mechanical properties of poly(D,L-lactic acid)/poly(e-caprolactone) blends. *Biomaterials* 24, 4181–4190, 2003.

39. (a) Kobayashi, S. and Sugimoto, S. Biodegradation and mechanical properties of poly(lactic acid)/poly(butylene succinate) blends. *Journal of Solid Mechanics and Materials Engineering* 2, 15–24, 2008; (b) Kim, Y.F., Choi, C.N., Kim, Y.D., Lee, K.Y. and Lee, M.S. Compatibilization of immiscible poly(l-lactide) and low density polyethelene blends. *Fibres and Polymers* 5, 270–274, 2004.

40. Chen, H., Pyda, M. and Cebe P. Non-isothermal crystallization of PET/PLA blends. *Thermochim Acta* 492, 61–66, 2009.

41. Lee, S. and Lee, J.W. Characterization and processing of biodegradable polymer blends of poly(lactic acid) with poly(butylenes succinate adipate). *Korea-Australia Rheology Journal* 17, 71–77, 2005.

42. Shishoo, R. The use of renewable resource based materials for technical textiles applications. In Miraftab, M. and Horrocks, A. R. (Eds.). *Ecotextiles – The Way Forward*

for Sustainable Development in Textiles. Cambridge, UK: Woodhead Publishing Ltd., 2007, pp. 109–127.

43. Reddy, N. and Yang, Y. Structure and properties of chicken feather barbs as natural protein fibres. *Journal of Polymers and the Environment* 15, 81–87, 2007.

44. Das, O., Bhattacharyya, D., Hui, D. and Lau, K.T. Mechanical and flammability characterisations of biochar/polypropylene biocomposites. *Composites Part B* 106, 120–128, 2016.

45. Ryan, Shannon, Christiansen, Eric L., Davis, Bruce A. and Ordonez, Erick *Mitigation of EMU Cut Glove Hazard From Micrometeoroid and Orbital Debris Impacts on ISS Handrails.* Houston, TX: USRA Lunar and Planetary Institute, NASA Johnson Space Center, 2009.

46. Karger-Kocsis, J. *Polypropylene: Structure, Blends and Composites.* London, UK: Chapman & Hall, 1995, Vol. 3.

47. Molna, P., Mitschang, R.P. and Felho, D. Improvement in bonding of functional elements with the fibre reinforced polymer structure by means of tailoring technology. *Journal of Composite Materials* 41(21), 2007, 2569–2583. doi: 10.1177/0021998307077379

6 Product Life Cycle Assessment and Suitability

6.1 INTRODUCTION

Life cycle management (LCM) establishes a relationship between life cycle partners to attain the maximum benefit from all technical products. Persuading partnership factors are standards, environment, regulations and constraints of economy. To attain best practices, the partners should have to cooperate and knew all parties at all life cycle stages. To minimize the risks and to secure the maximum result, each of them should be a part of the value-adding processes. The contribution made by current industrial manufacturing practices and consumption styles increase the manufacturer's responsibility.[1] Reducing environmental impacts throughout the life cycle of products while maintaining the company's position in the market and its place in society is called 'sustainable product design'.[2] Sustainable development 'meets the needs of present necessity without compromising the environmental gifts of future generations to meet their own needs'.[3]

As fears grow over climate change and the depletion of natural resources, it is essential to understand the environmental consequences of manufacturing. Life cycle assessment (LCA) is one of the techniques that can be used to address the issue. A systematic understanding of fibre-reinforced polymer composite and indeed the environmental impacts it has throughout its life cycle is needed, from its raw material stage to production, real-time usage, and end-of-life treatment method and process technique; this is known as life-cycle assessment – LCA-ISO, 2006. Such as assessment consists of four phases: (a) definition of goal and scope, (b) inventory analysis, (c) impact assessment, and (d) interpretation of results.[4]

Inventory investigation includes the compilation and quantification of input-to-output data for the composite product system throughout its life cycle. Two different approaches are followed in inventory analysis, namely process-based analysis and input-to-output–based analysis; both approaches come with their own strengths and limitations. Impact assessments have many derivatives, but it is preferable to consider environmental impacts like energy consumption and carbon dioxide emissions. The ISO 14040 series of standards provides detailed information on LCA.[4]

The LCA confirmed method involves analyzing and quantifying the environmental impacts of industrial production processes and products. As well as examining the ecological effects by determining the consumption of materials and energy and the resulting releases during the life of the product, this method helps to find tweak points and optimization opportunities. This kind of impact assessment strategy is useful for continuous product improvement.[5] Cumulated energy demand and resource consumption, as well as global warming potential (GWP), are the result of energy-related emissions. These potentially contribute to acidification and eutrophication in the relevant agricultural production chain, and this is considered in product life cycle impact assessment.[6]

Sustainable development proposes relationships between society and the environment; it requires fundamental changes in mentality and behaviour. Renewable materials will promote industrial systems and control toxic emits. 'Greener' composites would play a main role in future products, since they have environmental advantages such as reduced dependence on nonrenewable resources and the fact that they can disposed of as biodegradable material.[7,8] From groundwork to harvest, new jobs are being created in the production of fibre, including agricultural and processing job opportunities for local communities. The replacement of traditional materials such as wood, metals, minerals, and plastics in all sectors from construction to the computer industry benefits the environment.[9]

6.2 LIFE CYCLE ASSESSMENT SYSTEM

Product-oriented environmental policies aim to reduce the impact of products by adopting a range of measures:[10-14]

(a) Eco-labelling or supplying environmental information on products centred on LFA and reflecting the complete environmental impact of the product. Europe has several national eco-labelling schemes covering a range of different product types.
(b) Green public procurement, helping to guide public buyers on taking environmental considerations into account;
(c) Ranking of products based on their impact on the environment and view of resource availability to guide future regulation.
(d) Making manufacturers of certain product types, for example cars and electronics, responsible for taking back their products at the end of their useful life, thus relieving the customer of liability for the safe disposal of products.

Using an automatic colorimeter instrument (TCP2 – B, China) a composite's faded colour was measured according to the CIE L*a*b* colour system. The colour parameters' redness (a*), yellowness (b*). lightness (L*), and colour change (ΔE*) values were determined. After the weathering process, the composite became lighter (the colour faded) due to the depolymerization of cellulose and photodegradation of lignin, hemicelluloses, and other extractives present in bamboo fibre. Photodegradation of bamboo fibre generates carbonyl groups and hydroxy peroxide, photo-bleaching the composite's surface.[15] Hydroquinone formed through photodegradation of the matrices decreases the amount of yellowness, which would be the reason for the lighter colour of the composites after the weathering process.[16] A colour change of 35% in a bamboo fibre/polypropylene composite after 10 months of natural weathering has been shown to be near equal in value to that in a composite sample after 800 h of UV exposure.

Experiments were carried out on the composite specimen to determine physicomechanical properties such as density, which was measured as per ASTM D-792. Flexure composite bars for Type I tensile test, specimens measuring $150 \times 10 \times 4$ mm were acquired with a loading speed of 10 mm/min as per Chinese National Standard GB/T 1040-2006 procedure. The notched ISs of type IA test, specimens measuring $80 \times 10 \times 4$ mm, with 3 mm of remaining thickness specimen measuring were subjected to Chinese National Standard GB/T 1043-2008 procedure. Vicat softening points VST tests, the composite specimens with a diameter of 10 mm and a thickness of 4 mm considered for the test as per ASTM D-1525 procedure. Weather-exposed composite specimens and unexposed composite specimens are subjected to repeated test runs.[17,18]

An apparent decrease in composite mechanical properties and a slight reduction in density and VST was shown on composites that had been naturally exposed outdoors. Notched IS declined the most, then TS, while the smallest decline was observed in bending properties. Mostly, the degradation occurred on a thin surface layer of the composite, with the bulk of the specimens being comparatively unaffected. Moreover, bamboo fibres exerted a screening effect and efficiently improved the degradation resistance of composites. Compared to the UV-accelerated weathering, after 300, 600, and 900 h of UV exposure, the modulus of rupture retentions of the 35% bamboo fibre/PP foamed composite were 85.8%, 84.0%, and 81.6%; and the notched ISs retentions were 83.6%, 79.4%, and 77.1%, respectively. It was shown that 10 months of natural weathering had less effect on bending properties than 300 h of UV weathering, but the effect on the notched IS was greater than 900 h of UV weathering.[18]

Samples of composites measuring $10 \times 5 \times 5$ mm were subjected to ESEM (environmental scanning electron micrographs – XL30 PHILIPS, FET, the Netherlands), and the surfaces of un-weathered and weathered composites were examined. Due to photodegradation, cracks and voids formed in the composites. Swelling and shrinking caused by bamboo fibre absorption and desorption created a protruding fuzzy appearance on the composite surface.[18]

A rotational rheometer (Haake ARSIII, Germany) was used to measure complex viscosity, storage modulus, and loss modulus at a frequency range of 0.01–70 Hz at 195°C, 100 Pa. The 35% bamboo fibre/PP composite shows higher moduli than PP foam due to its greater stiffness. The consequence of shear stress on the linear viscoelastic part was characterized by the critical stress of the composite. In a stage of more than 100 Pa, stress, storage modulus G′ and loss modulus G″ for the unweathered composite was reduced. The composite subjected to natural exposure for up to 10 months showed decreases in each module. Critical stress is more uniform in a composite after natural weather exposure.[19]

XPS – ESCALAB 250, UK was used to determine the types of oxygen–carbon bonds present in the composite material. A value for the ratio of oxygenated to unoxygenated carbon (Cox/unox) for the weathered composite is higher than that for the unweathered one. The reduced C1 atomic ratio and amplified surface oxidation (Cox/unox) after weathering indicate lignin deterioration on the composite surfaces. The increased Cox/unox may be attributed to degradation of the matrices.[20] Life-cycle assessment of composites takes a life-cycle approach by considering all the facts for the real-time usage of the product, that is, its durability and its potential for reuse and recycling.

6.3 WEAKENING OF FIBRE COMPOSITE

An energy curve is defined as a loss in strain energy per cycle under fatigue loading in fibre-reinforced polymeric composite materials, and it usually comprises three stages. Stage I consists of 15–20% of the fatigue life of the material. Stage II consists of 70–75% of the life of the material and Stage III consists of 10% of life of the material.

Preliminary steep energy loss happens during Stage I and the number of cycles needed to reach the end of Stage I decreases with the amplified applied strain level. The rate of change of energy per cycle is called *energy release rate* (dU/dN). In Stage II, the energy curve is initiated as a constant and characteristic of the constitutive material for a particular strain level.

The energy release rate (dU/dN) differs from the applied strain level by a power function and the coefficients of the power called *fatigue coefficients* (a, b). It can be used to predict the useful life of fibre-reinforced polymeric composites under fatigue loading. The fatigue resistance of the composite materials increases with higher values of the fatigue coefficient (b). A 50% increase in expended strain energy at the end of Stage II is commonly observed in composite test specimens. Stage III is characterized by a steep energy loss curve per cycle, leading to the failure of a composite test specimen. Henceforth, the end of Stage II is assumed as the fatigue life of the material so as to arrive at a conformist approach for the fatigue life prediction of fibre reinforced composite material. The following factors are taken into account in determining the fatigue resistance of FRP composite materials:

- Nature and type of fibre
- Fibre volume fraction and loading direction
- Percentage of fibre alignment in 0°, 90°, and 45°
- Thickness of the composite material (0.5 in. or less for 2D stitched fabrics)
- Mode of loading

The experimental and predicted fatigue that exists at numerous strain levels are compared in S–N curves and the model is found to be conservative with 10% error. A simultaneous fibre failure (SFF) model is used to predict the long-term durability of fibre composite materials.[21]

6.4 NATURAL WEATHERING

Natural weathering tests are usually done to determine the durability of fibre-reinforced composite material under natural circumstances. The purpose of the experiments is to manipulate and determine the amount of biodegradability possessed by the fibre reinforced composite while exposed to natural weather. Rain, sunlight, dew, and wind are common facets of natural weathering. Ultraviolet (UV) rays from sunlight are reportedly the most common cause of polymer object failure in the course of natural weathering. Absorption of UV rays with moisture on the surface of fibre-reinforced polymer composite creates free radicals. These free radicals react with oxygen and degrade the unsaturated polymer during natural weathering. Cracking, chalking, yellowing, brittleness, and loss of transparency are commonly seen in conventional natural fibre reinforced polymers exposed to natural weather.[22-25]

The natural weathering test is performed on composites by placing them on a rooftop, with the sample facing from north to south at an inclined angle. It was left for a period of six months. The recorded average weather condition in Tiruppur, India, for the duration of exposure is presented in Table 6.1. The exposed samples were retrieved and their thermal conductivity, sound absorption, and mechanical properties were measured. Acrylic-based silicone emulsion (ASE) coated samples and uncoated samples were considered for analysis. The knife-blade coating method is adopted for the application of acrylic-based silicone emulsion coating[26] (Figures 6.1 and 6.2).

As seen in Table 6.2, before the natural weathering process, samples D, A, C, and B show the lowest order of thermal conductivity, respectively. Structural differences play an important role in thermal conductivity. After the natural weathering process, due to the structural weakening of the composite, cracks and porosity increase, improving the thermal conductivity value of the fibre. When compared to the flexible structure D, the structures A, B and C are not further weakened by the natural weathering process. The loose structure of D was not bounded by any hot compression bonding, and since it had more holes in its structure, sunlight and other natural weathering factors easily penetrated into it.[26]

Jute fibre contains lignin, which is highly sensitive to sunlight and forms free radicals. Both natural and synthetic polymers absorb UV radiation and undergo thermo-oxidative, photo-oxidative, and photolytic reactions. This degrades the surface abrasion of composites and reduces their toughness. After the weathering process, without ASE coating, the thermal conductivity value of composites is increased, whereas, ASE-coated composites show good thermal insulation values.

TABLE 6.1

Average weather condition (accuweather; worldweatheronline; weather.com)[26]

Natural weathering: Weather criteria	Scale level
High temperature	34°C
Low temperature	22°C
Wind	5 mph ENE
Rain	2–3 mm
Precipitation	4.3 mm
Humidity	51%
Pressure	1009 mbar
Cloud cover	49%
UV index	5–8

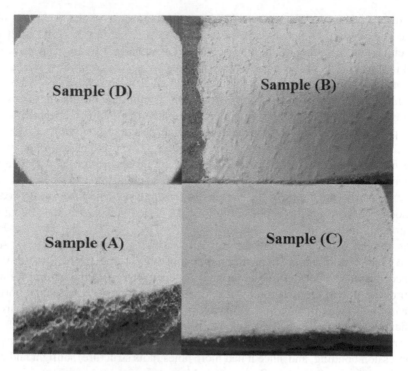

FIGURE 6.1   Before weathering ASE coated composites.

FIGURE 6.2   Before weathering ASE coated composites.

TABLE 6.2
Thermal conductivity value of composites[26]

| Sample | Density (kg/m³) | Thickness (mm) | Thermal conductivity (λ) before weathering | | Thermal conductivity (λ) after weathering | |
			Without ASE coating (W/mK)	With ASE coating (W/mK)	Without ASE coating (W/mK)	With ASE coating (W/mK)
A	486	6.75	0.0201	0.0199	0.0400	0.0200
B	592	5.54	0.0259	0.0214	0.0359	0.0227
C	524	6.25	0.0213	0.0201	0.0222	0.0202
D	461	7.12	0.0197	0.0190	0.0412	0.0290

Before weathering, ASE coated composites show reduced thermal conductivity value compared to uncoated composites.[26]

The application of ASE coating to the surface of composites reduces the thermal conductivity value of composites except for structure D. The porous structure of composites is blocked by the ASE coating. The uneven surface structure of D was not covered completely, yet penetration of weathering factors affected the structure and improved the porosity, causing thermal conductivity to be slightly reduced compared to without ASE weathering. Samples A, B, and C show less thermal conductivity value with ASE coating after the natural weathering process.[26]

As seen in Table 6.3, before the natural weathering process, samples A, D, C and B show the highest order of average noise reduction coefficient respectively, at low frequencies from 250 Hz to 2000 Hz frequency level. Material pore size decides the efficiency of acoustic absorbency. If the pore is too big, the sound passes through it, whereas if it is too small it may act as an inefficient acoustic absorber. Solid fibre-based structures increase internal reflection and reduce the energy level in the structure through frictional losses.[26]

Before and after the weathering process, the ASE-coated composites show a reduction in average noise absorption when compared to an uncoated sample. The

TABLE 6.3
Noise reduction coefficient value of composites

| | Noise reduction coefficient (%) | | | |
| | Before weathering | | After weathering | |
Sample	Without ASE coating	With ASE coating	Without ASE coating	With ASE coating
A	0.64	0.54	0.68	0.52
B	0.57	0.46	0.60	0.48
C	0.59	0.47	0.60	0.45
D	0.69	0.58	0.75	0.56

uncoated sample weakens after weathering. The process improves the porous structure, abrades the surface structure, and improves the sound absorbency. In the coated sample, however, the porous surface is blocked by the ASE, reducing the absorbency level before and after weathering. Sound transmission can be reduced by decoupling (creating air gaps or air spaces between two partitions) and by adding mass to the material. Blocking of sound attained through solid material may resist the transmission of sound waves. The ASE-coated composite material after weathering may work as a good blocking material and may also act as a low-absorbing medium. A new way of reducing sound by replacing reflection with dissipation was developed.[26]

Before weathering, the ASE coated and uncoated samples do not show differences in their mechanical properties. After weathering, however, the mechanical properties of composites were slightly affected due to residual stresses. When composites were exposed to the weathering process, the deterioration was less when compared to the pure polymer material. A slight reduction in mechanical properties occurred in the sample without ASE coating after weathering. Due to its loose bonded structure and the absence of compressed hot pressing, sample D was affected more when compared to the rest of the samples.[26]

Sample C with added low-melt polyester, being a rigid composite, is a tough surface showing good mechanical properties both before and after exposure to natural weathering. The mechanical properties improved slightly in the ASE coated Sample C after weathering. The uncoated Sample C after weathering was slightly affected.[26]

The rigid Sample C showed high flexure modulus, flexural strength, and strain at break. Next to that, rigid Sample B showed average strength. Samples A > D being flexible showed less flexural properties. Mechanical properties are important when considering composites for use in indoor and outdoor applications. Especially after the natural weathering process, the jute and HCP composite with 5% low-melt polyester (Sample C) shows excellent mechanical characteristics.[26]

As seen in Table 6.4, the limited oxygen index value for coated and uncoated composite runs from highest to lowest thus: Samples C, B, A and D. Rigid and tough structure, C, had less trapped air and thus required a high level of oxygen for its combustion. By following the order D, A, and B, trapped air content present in the composite structure comparatively requires less oxygen for its combustion.[26]

Less than 20.94% of the limiting oxygen index (LOI) value of the material will burn easily in air. The jute/HCP composites meet more than the minimum

TABLE 6.4
LOI value of composites[26]

		LOI %	
Sample	Composite type	Uncoated composites	ASE-coated composites
A	Rigid	23	24
B	Rigid	26	26
C	Rigid	28	28
D	Flexible	22	24

flame-retardant level. The right kind of organic polymer coating may delay ignition and prevent flame spread. Here the ASE coating is intentionally applied for weather-proofing, not for any other specific end use.[26]

6.5 ARTIFICIAL WEATHERING

Accelerated weathering simulates the destructive effects of long-term outdoor expo-sure of composite materials by exposing them to varying conditions of the most aggressive weathering components such as heat, light, and moisture. A xenon arc light and carbon light source are used to provide a radiation spectrum that simulates natural sunlight.

Azuma et al. performed the accelerated weathering test on polypropylene com-posite at 63°C using different kinds of light sources like metal halide – IWASAKI Electric, EYE Super UV-W151, Xenon – SUGA test, Super Xenon Weather Meter SX75 and carbon arc – SUGA test, Sunshine Weather Meter WEL-SUN-DC-B lamps. The ISO 4892-2 procedure was followed, with daylight filters of 275 nm wavelength. In the xenon lamp test, a 120 h radiation exposure and 18 min simul-taneous water spray of 480 ml/min was utilized. In the carbon arc lamp test, a 60 h radiation exposure and 12 min simultaneous water spray of water of 2.1 ml/min was utilized. In the metal halide lamp test, a dew condensation water cycle method used with a 6 h cycle, 5 h radiation exposure, and 1 h dew condensation. After the testing process, the order of degradation rate of fibre reinforced composite expressed for the outdoor exposure was shown to be higher in Xenon lamp test; the next highest level of degradation was shown in the metal halide lamp test, and finally a small amount of degradation was shown in the carbon arc lamp test.[27]

6.6 STRENGTHENING COMPOSITE PERFORMANCE

Durability of the composite depends on fibre – matrix volume fraction, relevant to this taking care of kenaf fibre volume fraction not reached minimum value. It ensures fibre-controlled composite and does not provide the way to matrix dominated fail-ures.[28,29] Sometimes, the presence of voids in a fibre-reinforced composite structure might be a contributing cause for a drop in tensile strength at a lower volume frac-tion of kenaf fibre.[30,31] Commonly, in fibre-reinforced composite, maximum strain is reduced at a higher fibre volume fraction content. Moisture behaviour, that is, sorp-tion and desorption affect the mechanical behaviour of the composite. Water par-ticles act as plasticizers in the composite material, increasing the maximum strain of the composite after water absorption.[32]

The density and thickness of the composite material play a significant role in the moisture absorption behaviour of fibre-reinforced composite material. In preform manufacturing such as woven, knit, and nonwoven fabrication processes, the cover factor value needs to be analyzed based on the end-use application of the composite material. Instead of resin, thermoplastic or thermoset fibres and other additives are blended together to form preforms. Random and uniform distribution of matrix fibre is incorporated with the core fibre during compression moulding or any other heat-setting process.

Appropriate selection of fibre, preprocessing of fibre, compatibility of fibre, structure and float choice of warp and weft yarn in woven fabrication, dial feed yarn or cylinder feed yarn or alternate feed of yarn and similar choice of yarn selection in knit fabrication, needling density, blending method (air laid, cross laid, multiple cross laid), choice of nonwoven fabric production, and many other critical factors need to be considered when producing high-performance composite structures.

Acrylic-based emulsion coatings, fire-retardant finishes, water-repellent and oil-repellent finishes, and other kinds of paints and coatings help to protect the surface of fibre-reinforced composites. The coated finishes give protection from dew, sunlight, moisture, fire, dirt, and dust. Surface protection need not allow the external factors to get inside of the composites and degrade the internal structure of the composite, so that right and appropriate selection of surface finishes over composite is preferred for the long term durability. It increases the life of the product and improves its functional performance.

6.7 CONCLUSION

Improvement of the performance of composites will be attained through research on structural components, geometry prediction, and its associate effects. The cost of manufacturing can be reduced by adopting natural fibre for use in composite production. Natural fibres in composites have some fascinating characteristics compared to synthetic fibres. Low fossil-fuel energy requirements, the availability of natural fibre in huge quantities, its cost effectiveness, its renewability, and its reasonably good mechanical properties are a few examples. The moisture-uptake property of a composite makes a significant impact on its mechanical properties. Water absorption behaviour of composites in conditions of relative humidity radically affect their mechanical properties.[33] The effects of long-term weathering on materials can be predicted over a short period by using an artificially accelerated weathering test. Actually, a plastic composite does not exhibit exactly the same reaction in actual use over a period of time. Fibre-reinforced composites with unavoidable voids along with appropriate stabilized coating show best weathering resistance and durability. A jute fibre–reinforced composite door frame satisfactorily meets the IS: 4021–83 standard and shows a satisfactory life cycle; even after 3.5 years, it does not show any signs of cracks, holes, or dimensional instability.[34]

REFERENCES

1. Alting, L. and Legarth, J.B. Life cycle engineering and design. *CIRP Annals* 44(2), 569–580, 1995.
2. Seamann, R. *Environmental Strategies, Ethics and Management.* Sweden: Swiss Academy of Engineering Sciences, CAETS Convocation, 1995.
3. Our Common Future (The Brundtland Report), 1987 (Oxford University Press).
4. Zhang, C. Life cycle assessment (LCA) of fibre reinforced polymer (FRP) composites in civil applications. Life Cycle Assessment (LCA), Eco-Labelling and Case Studies, 565–591, 2014. doi: 10.1533/9780857097729.3.565
5. Wenzel, H., Hauschild, M.Z. and Alting, L. Environmental assessment of products. In *Volume 1: Methodology, Tools and Case Studies in Product Development.* London, UK: Springer, 1997.

6. Flake, M., Fleissner, T. and Hansen, A. Ecological assessment of natural fibre rein-forced components and thermoplastics for automotive parts. *Progress in Rubber, Plastics and Recycling Technology* 18(4), 219–245, 2002.
7. Alves, C., Dias, A.P.S., Diogo, A.C., Ferrão, P.M.C., Luz, S.M., Silva, A.J., Reis, L. and Freitas, M. Eco-composite: The effects of the jute fiber treatments on the mechani-cal and environmental performance of the composite materials. *Journal of Composite Materials* 45(5), 573–589, 2010.
8. Rowell, R.M. Potentials for jute based composites. In Jute India '97, Expo-Business Meet-Seminar, Pragati Maidan, New Delhi, 1997.
9. Harish, S., Peter, M.D, Bensely, A., Mohan, L.D. and Rajadurai, A. Mechanical property evaluation of natural fiber coir composite. *Materials Characterization* 6, 44–49, 2009.
10. Frees, N. and Pedersen, M.A. *To Choose or To Lose*. The Hague: National Environmental Policy Plan, Ministry of Housing, Physical Planning and Environment, 1989.
11. *A Strengthened Product-Oriented Environmental effort (in Danish)*. Copenhagen: Danish Environmental Protection Agency, 1996.
12. *Account of the Danish Environmental Protection Agency on the Product-Oriented Environmental effort (in Danish)*. Copenhagen: Danish Environmental Protection Agency, 1998.
13. Scholl, G. Sustainable product policy in Europe. *European Environment* 6, 183–193, 1996.
14. European Commission Workshop on Integrated Product Policy, Brussels, 8 December 1998, Final Report (Directorate- General XI, Environment, Nuclear Safety and Civil Protection).
15. Butylina, S., Hyvärinen, M. and Kärki, T. A study of surface changes of wood-poly-propylene composites as the result of exterior weathering. *Polymer Degradation and Stability* 97(3), 337–345, 2012.
16. Fabiyi, J.S. and McDonald, A.G. Wood plastic composites weathering: Visual appear-ance and chemical changes. *Polymer Degradation and Stability* 93(8), 1405–1414, 2008.
17. Zhou, X.X., Chen, L.H., and Huang, S.S., Accelerated weathering of bamboo flour/polypropylene foamed composite. *Trans CSAE* 30(7) 287–292, 2014.
18. Zhou, X., Huang, S., Yu, Y., Li, J. and Chen, L. Outdoor natural weathering of bam-boo flour/polypropylene foamed composites. *Journal of Reinforced Plastics and Composites* 33(19), 1835–1846, 2014.
19. Lion, A. and Kardelky, C. The Payne effect in finite viscoelasticity: Constitutive mod-elling based on fractional derivatives and intrinsic time scales. *International Journal of Plasticity* 20(7), 1313–1345, 2004.
20. Matuana, L.M., Jin, S. and Stark, N.M. Ultraviolet weathering of HDPE/wood-flour composites coextruded with a clear HDPE cap layer. *Polymer Degradation and Stability* 96(1), 97–106, 2011.
21. Natarajan, V., Gangarao, H.V. and Shekar, V. Fatigue response of fabric-reinforced polymeric composites. *Journal of Composite Materials* 39(17), 1541–1559, 2005.
22. Stark, N.M. and Matuana, L.M. Influence of photo stabilizers on wood flour-HDPE composites exposed to xenon-arc radiation with and without water spray. *Polymer Degradation and Stability* 91, 3048–3056, 2006.
23. Schier, J. *Compositional and Failure Analysis of Polyesters: A Practical Approach*. New York: John Wiley and Sons, 2000.
24. Copinet, A., Bertrand, C., Govindin, S., Coma, V. and Couturier Y. Effects of ultravio-let light (315 nm), temperature and relative humidity on the degradation of polylactic acid plastic films. *Chemosphere* 55, 763–773, 2004.
25. Sharma, S.C., Krishna, M., Narasimhamurthy, H.N. and Sanjeevamurthy. Studies on the weathering behaviour of glass coir polypropylene composites. *Journal of Reinforced Plastics and Composites* 25, 925–932, 2006.

26. Zakriya, G.M. and Ramakrishnan, G. Jute and Hollow Conjugated polyester composites for outdoor & indoor insulation applications. *Journal of Natural Fibres* 16(2), 185–198, 2019. doi: 10.1080/15440478.2017.1410515

27. Azuma, Y., Takeda, H., Watanabe, S. and Nakatani, H. Outdoor and accelerated weathering tests for polypropylene and polypropylene/talc composites: A comparative study of their weathering behavior. *Polymer Degradation and Stability* 94, 2267–2274, 2009.

28. Huda, M.S., Drzal, L.T., Mohanty, A.K. and Misra M. Chopped glass and recycled newspaper as reinforcement in injection molded poly(lactic acid) (PLA) composites: A comparative study. *Composites Science and Technology* 66, 1813–1824, 2006.

29. Iannace, S., Nocilla, G. and Nicolais, L. Biocomposites based on sea algae fibres and biodegradable thermoplastic matrices. *Journal of Applied Polymer Science* 73, 583–592, 1999.

30. Hill, C.A.S., Abdul, H.P.S. and Khalil. The effect of environmental exposure upon the mechanical properties of coir or oil palm fiber reinforced composites. *Journal of Applied Polymer Science* 77, 1322–1330, 2000.

31. Facca, A.G., Kortschot, M.T. and Yan, N. Predicting the elastic modulus of natural fibre reinforced thermoplastics. *Composites Part A: Applied Science and Manufacturing* 37, 1660–1671, 2006.

32. Joseph, P.V. Environmental effects on the degradation behaviour of sisal fibre reinforced polypropylene composites. *Composites Science and Technology* 62, 1357–1372, 2002.

33. Rashdi, A.A.A., Sapuan, S.M., Ahmad M.M.H.M. and Khalina, A. The effects of weathering on mechanical properties of kenaf unsaturated polyester composites (KFUPC). *Polymers & Polymer Composites* 18(6), 337–343, 2010.

34. Singh, B. and Gupta, M. 2005. Performance of pultruded jute fibre reinforced phenolic composites as building materials for door frame. *Journal of Polymers and the Environment* 13(2), 127–137, 2005.

7 Testing of Composites

7.1 INTRODUCTION

Composite materials sometimes suffer effects during the manufacturing process itself, called *defects*. These include voids being filled with volatile resin components, defects in fibre-to-resin bonding, delamination of plies, ply cracks, external impurities, and so on. In-service defects can also occur, like cracks, fracture of fibres, impact damage, ingress of moisture, weathering effects, and so on. Defects in composite materials due to manufacturing faults may endanger human life and create huge losses when it comes to commercialization. It is mandatory to analyze the quality level and inspect for manufacturing defects, particularly subsurface defects, in composite materials before they come into final use.[1]

The quality of fibre-reinforced composite material can be determined using a graph theoretic approach. The fibre-reinforced composite material coming out from manufacturing processes should have a good quality surface finish without any defects. Several factors determine the quality, from the design to manufacturing of composites using the resin transfer moulding process. Many factors are used to produce a quality digraph based on the relations existing between them. Hence, the quality of digraph depends on defining the fuzzy crisp scores for the degree of interactions between the factors. A quality index is derived from a quality stable function. Its help the quality inspectors on the shop floor to decide whether the product is within benchmark limits. An upper and lower limit bound in deriving the indices proposes a quality index. This index is a useful numerical index for quality assurance staff to take proper decisions at shop-floor level on the manufacturing process.[2]

Infrared (IR) thermography is an emerging convenient method for the nondestructive testing of composite materials. This technique is reliable, fast, and allows real-time measurement over a large surface area. Thermography provides global mapping whereas other methods like ultrasonic testing will give only localized results. Pulsed and lock-in thermography are the most frequently used nondestructive assessment techniques. In pulsed thermography, the composite specimen is heated using a short energy pulse in the form of light. The thermal response of the composite material is recorded by an IR camera. The subsequent thermal image from the IR camera reveals defects present in the composite material at different depths. In lock-in thermography, the composite material is subjected to single-frequency sinusoidal thermal excitation. The magnitude and phase of the reflected thermal wave within the composite material is consequently recorded as thermal images, revealing any subsurface defects in the composite material.[3-5]

7.2 PRINCIPLE OF FREQUENCY-MODULATED THERMAL WAVE IMAGING

In active thermography, the composite material is exposed to an external energy source, and its reaction is recorded after a determined time. The thermal wave produced at the surface propagates inside the composite material by diffusion. The wave reflects damage or defects such as voids, delamination, cracks, and so on. Diffusion rate of composite is mostly influenced by its defects so that temperature difference exists in the damaged part of the composite compared to the non-damaged part. All the interference of received and reflected waves produces objects on the surfaces. A harmonic fluctuates radiation pattern detected by an IR camera. The experimental arrangement is shown in Figure 7.1. The sample is intermittently heated using tungsten–halogen flood lamps.[5]

The Zenoptik long-wave IR camera has a range of sensitivity from 7.5 to 14 μm. It has a maximum frame speed of 50 Hz with a pixel resolution of 640×480 pixels. Energy is focused on the surface of an opaque material, and the composite material absorbs some of the incident energy. These processes induce a localized heat flow in the composite specimen. This is clearly depicted in Figure 7.1.

Time-dependent heat flow is ruled by the one-dimensional heat diffusion equation where T is temperature and α is thermal diffusivity $\alpha = k/\rho$; k is mass density, ρ specific heat of the medium respectively (Figure 7.2).

$$\frac{\partial^2 T(x,t)}{\partial x^2} = \frac{1}{\alpha} \frac{\partial T(x,t)}{\partial t} \tag{7.1}$$

Artificially produced subsurface defects in fibre-reinforced composite at different depths are detected and applied using an image reconstruction algorithm. The defects are captured as amplitude and phase images, with phase images being considered more reliable. The observation of defects at slower excitation frequency a gives greater depth of result. It is an option for nondestructive testing of natural fibre–reinforced composites, offering reliable, safe, and cost-effective assessments.[5]

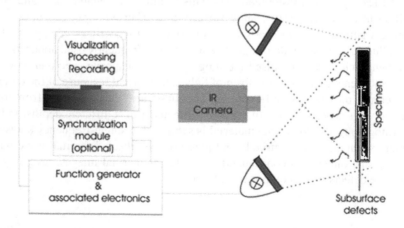

FIGURE 7.1 Investigational arrangement of nondestructive testing using thermography.[5]

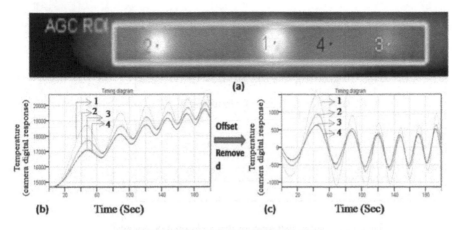

Surface temperature response of the composite

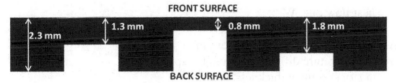

Artificially generated defects in composite sample

FIGURE 7.2 Frequency-modulated thermal wave imaging.

7.3 PHYSICAL TESTING

7.3.1 MEASUREMENT OF THERMAL CONDUCTIVITY AND ITS RELATED DERIVATIVES

According to ASTM C518, a steady-state thermal conductivity (λ) property was determined by a NETZSCH HFM 436 Lambda model heat flow meter. A sample size of 30.5 cm × 30.5 cm was used at a temperature of 90°C on the upper plate and 30°C on the lower plate. The inferior plate association is transportable; hence, composite samples of varying thickness and different composition percentages of fibre can be tested.[6]

Measuring the heat flow in a down-structure direction, as shown in Figure 7.3, minimizes the convective heat transmission from end to end by the permeable insulation materials. Heat transfer in the heat-flow meter instrument mainly occurs due to solid/gas conduction and thermal radiation mechanisms. The plate temperatures are controlled by bidirectional heating and cooling Peltier systems, which are attached with a closed-loop fluid flow with an incorporated forced air heat exchanger. Using two heat-lux temperature sensor transducers, the temperature was measured according to a time limit maintained at 12–14 min for the prepared sample.[6] The unit measurement of thermal conductivity (λ) is W/(m K).

$$Q = \lambda A \frac{\Delta T}{L} \, \mathrm{W}/(\mathrm{m} \cdot \mathrm{K}) \tag{7.2}$$

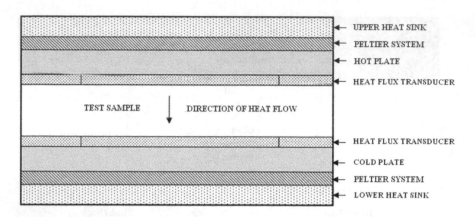

FIGURE 7.3   Heat flow meter schematic for thermal conductivity testing.

where

Q = heat flow rate, W
λ = thermal conductivity, W/(m.K)
A = meter area normal to heat flow, m²
ΔT = temperature difference across the specimen, K
L = in-situ specimen thickness, m

According to EN ISO 6946, the thermal resistance achieved for a constructional layered composite was obtained by dividing the thickness (T) (Polyurethane Foam Associations 2006) by the thermal conductivity value of a nonwoven composite:

$$R = T / \lambda \ \left(m^2 K \right) / W \tag{7.3}$$

The thermal transmittance (U) is the time rate of heat flow through 1 m² of a composite component when the temperature difference between the surfaces in the direction of heat flow is 1 K. The U value can be calculated for a given composite (BS 6993 1989) and is generally represented in W/(m2K).

$$U = 1 / R \, W \Big/ \left(m^2 K \right) \tag{7.4}$$

Thermal diffusivity (α) measures the capacity of a composite to conduct thermal energy compared to its capacity to store thermal energy. It is an indicator of how rapidly a composite will vary in temperature in reaction to the relevance of heat. It is usually denoted by α in m²/s.

$$\alpha = \frac{\lambda}{\rho C_p} \ m^2 / s \tag{7.5}$$

where

λ = thermal conductivity W/(mK)
ρ = density (kg/m³)
Cp = specific heat capacity J/(kgK)

7.3.2 Measurement of Air Permeability

Air permeability testing can be used to provide an indication of the breathability of weather-resistant, rainproof fabrics and coated fabrics. In general, it is used to detect changes during the manufacturing process. Constructional elements and finishing techniques can have a considerable effect upon air permeability by affecting a change in the length of airflow paths through a fabric. Hot calendaring will be used to smooth fabric components, thus decreasing air permeability. Fabrics with different kinds of apparent textures on either side can have a different air permeability depending on the direction of air flow (ASTM D737).[7]

A steady flow of air is introduced perpendicularly through the test area and the air-flow rate is adjusted to preferably provide pressure differentials of between 100 and 2500 Pa (10 and 250 mm or 0.4 and 10 in. of water) between the two surfaces of the fabric being tested. At a minimum level, the test apparatus must provide a pressure drop of 125 Pa (12.7 mm or 0.5 in. of water) across the specimen (ASTM D737).[7]

The test specimen was placed on the test head of the instrument, and the test was performed as specified in the manufacturer's operating instructions; it was continued for ten specimens of each sample. The air permeability was determined in accordance with ASTM D737.

7.3.3 Measurement of Breaking Force and Elongation

ASTM D3039 tensile testing is used to measure the force required to break a fibre-reinforced composite specimen and the extent to which the specimen stretches or elongates to that breaking point. Specimens are placed in the grips of a universal test machine at a specified grip separation and pulled until failure. For ASTM D3039, the test speed can be fixed by the material property or time to failure (1–10 min). A typical test speed for standard test specimen is 2 mm/min (0.05 in/min). An extensometer or strain gauge is used to determine elongation. Specimen size is a constant rectangular cross section, which is 25 mm wide and 250 mm long (ASTM D3039).[7]

7.3.4 Measurement of Sound Absorption Coefficient

The sound absorption coefficient of jute/HCP composites was tested using an impedance tube method based on ASTM E1050. Sound wave energy was created and passed through the composite material, directed by the fibres. Along the flow path, some of the energy was partially absorbed, reflected and converted into heat. The sound absorption coefficient indicates how much of the sound is absorbed by the composite material. It is expressed in a value between 1 and 0; perfect absorption, 1, indicates no reflection in the composite material and 0 indicates total reflection occurred in the composite material.[8] The absorption coefficient can be expressed as

$$\alpha = \frac{I_a}{I_i} \tag{7.6}$$

where

   α = Sound absorption coefficient

   I_a = Sound intensity absorbed (W/m^2)

   I_i = Incident sound intensity (W/m^2)

At one end of the impedance tube, a loudspeaker is attached. It produces broadband sound waves randomly. The composite test sample was placed at the other end of the tube. It receives the propagated plane wave generated by the loudspeaker. The plane wave is partially absorbed and subsequently reflected by a composite sample. The acoustical properties of the composite sample were tested in the frequency ranges from 63 to 6300 Hz.[8]

Up to a frequency of 2000 Hz, a 29 mm tube diameter is used, while at levels of more than 2000 Hz, a 100 mm diameter tube is used. Digital frequency analysis system and microphone tests are used to determine the sound absorptive potential of composites, with results reported for various frequencies.[8]

Based on ASTM E1050, the average sound absorption coefficient of jute/HCP composites was tested using the impedance tube method at frequency levels of between 250 Hz and 2000 Hz.[8]

7.3.5 MEASUREMENT OF ELECTRICAL RESISTANCE

The electrical resistance (R) property of composites according to the DIN54 345 standard is determined by Fischer Elektronik, Milli-TO 3 model ohm and current meter instrument under the following conditions: voltage: 250; current: 1.6 mA at 10 kΩ load resistance; temperature: 23°C; and relative humidity: 25%. Absorption current in an insulated composite material can be circulated by surface resistivity and volume resistivity. Resistance across the surface of a composite material in contact with the electrodes is considered for surface resistivity. The length, width, and thickness of composites are considered for volume resistivity.[8]

7.3.6 MEASUREMENT OF TENSILE TESTING

ASTM D3039 tensile testing is used to measure the force required to break a fibre reinforced composite specimen and the extent to which the specimen stretches or elongates to that breaking point. Specimens are placed in the grips of a universal test machine at a specified grip separation and pulled until failure. For ASTM D3039, the test speed can be determined by the material specification or time to failure (1–10 min). A typical test speed for a standard test specimen is 2 mm/min (0.05 in/min). An extensometer or strain gauge is used to determine elongation. Specimen size is a constant rectangular cross section, which is 25 mm wide and 250 mm long (ASTM D3039 2017). A 10 kN load cell is used to determine the tensile modulus of the composites.[8]

7.3.7 MEASUREMENT OF ROCKWELL HARDNESS NUMBER

ASTM D785–03 is used to measure the hardness number of composite material. A Rockwell hardness number is directly related to the indentation hardness of a composite material; the higher the reading, the harder the material (ASTM D785 2003).[8] The Rockwell hardness number is derived from the following relationship:

$$HR = 150 - e \qquad (7.7)$$

where,

HR = the Rockwell hardness number, and

e = the depth of impression after removal of the major load, in units of 0.002 mm.

7.3.8   MEASUREMENT OF CHARPY IMPACT TEST

A Charpy impact test on a notched jute/HCP composite was done according to ASTM D6110 using a universal impact testing machine. This is used to determine the resistance of composites when a notched sample of the composite is rested on a horizontal simple beam and is broken by the impact energy of a single swing of a pendulum. The energy lost by the pendulum during the breakage of the composite is equal to the sum of energies required to initiate fracture of the composites. The result obtained from this test is reported in terms of energy absorbed per unit of composite width (ASTM D6110 2004).[8]

7.3.9   MEASUREMENT OF THREE-POINT BENDING TEST

A three-point bending test was carried out on a Zwick Roell Z010 model universal testing machine using a 10kN load cell, according to ASTM D 790–02 standard. The sample was deflected up until a break occurs on the outward surface of the sample or up to a maximum strain of 5% is reached. The strain rate was maintained at 0.01 mm/mm/min. The sample is horizontally placed on two point contacts of a lower span and a force is applied from the top of the sample through one point of contact, leading the sample to bend or break (ASTM D790 2017). An average of 10 readings was noted for accuracy purposes.[9]

7.4   CHEMICAL TESTING

Kenaf and hemp fibres were treated using a 5% solution of sodium hydroxide for 1 h at room temperature, rinsed with tap water, and then neutralized in a 2% glacial acetic acid and tap water solution. The fibres were rinsed again with deionized water and dried at room temperature for 24 h. The fibres were treated using a 1% solution of 3-glycidoxypropyltrimethoxysilane in deionized water and ethanol in a 1:1 ratio. The pH of the solution was adjusted to 4 using a 2% glacial acetic acid. The solution was stirred for 2 h before the fibres were added. The fibres were washed with deionized water after the silane treatment. The fibres were then air dried for 12 h and then cured in a convection oven for 5 h. An epoxy was mixed with 15% weight of fibres and the mixture was degassed in a 373.15 K vacuum oven for 20 min. The composites were cured in a 418.15 K convection oven for 2 h. An alkali treatment eradicates lignin and hemicellulose contents from the surface of natural fibres. Alkali-treated kenaf and hemp fibre composites absorb more water. Alkali and silane-treated composites absorb less water. Kenaf composites absorbed less water compare to hemp fibre composites.[10] These phenomena are clearly depicted in Figure 7.4.

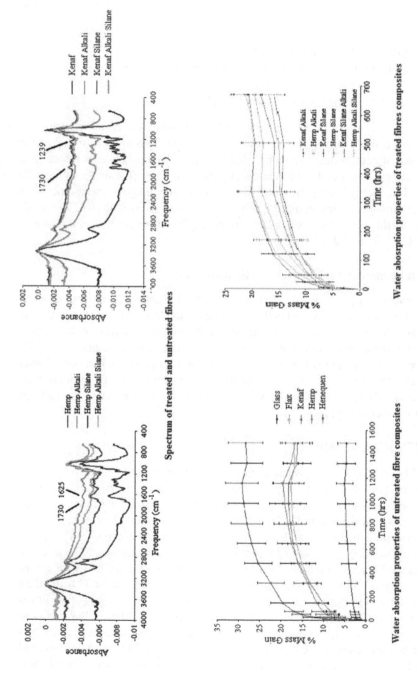

FIGURE 7.4 Alkali and silane treated and untreated composites absorption properties.[10]

7.4.1   Measurement of Limiting Oxygen Index

According to standard ASTM D2863, testing was used to find the minimum percentage of oxygen required to marginally support candle-like combustion of composites in a glowing mixture of oxygen and nitrogen in a test atmosphere. The composite sample was placed vertically in a chimney made of glass, and an oxygen/nitrogen environment was customized with a flow from the bottom of the chimney. The top edge of the composite sample was ignited, and the oxygen concentration in the flow was reduced little by little until the flame was no longer supported. The composite sample size was 80–150 mm long and 10 mm wide and was kept in a vacuum oven for drying at 70°C for 48 h. The results obtained from this test method cannot be used to determine the fire hazards of the composite sample in other fire circumstances (ASTM D2863 2017).[9] The Limiting Oxygen Index (LOI) was evaluated with the following equation

$$LOI = \frac{[O_2]}{[O_2]+[N_2]}\%$$
(7.8)

7.5   BIOLOGICAL TESTING

7.5.1   In Vitro Testing

Liana Cont et al. carried out experiments on PLA composites relevant to tissue engineering. They conducted cell morphology at four circumstances: developed one, poly (dopamine) functionalized, biotin functionalized and pressed PLA at different timing 3 hr, 17hr and 24 hr of incubation. After 3 hr, cells begin to elongate along the fibres and make bridges between the structure's pores through pseudopodia. Due to the large gaps between fibres, the cells begin infiltrating the fibrous scaffold. No differences can be seen in cell morphology and density on biotin functionalized fibres after 3 hr. The polydopamine functionalized fibres show more elongated cells due to dopamine's ability to modify the polymer's hydrophobic performance into a more hydrophilic one, facilitating cell adhesion. The pressed PLA results in cells with a more spread aspect, with gravity and sedimentation bringing cells to the surface.[11]

After 17 hr, cells are much denser and start dividing and spreading over the whole surface. The sample grows to 30 mm in size with a fibre diameter of 6 mm; fibroblasts tend to grow along the fibres, binding fibres together and wrapping around the thicker ones.[11] Cells continue to creating bridges between the fibres and to spread densely for up to 24 hr. Cell viability and metabolic activity were investigated by Alamar Blue assay after 1, 3, and 7 days of incubation, as clearly shown in Figure 7.5. The Alamar blue results after all the tested periods 1, 3, and 7 days of incubation shows significant differences ($p < 0.05$) between the tissue culture plastic control and all the samples being used.[11]

Cell propagation was examined by the DNA Hoechst method after 1, 3, and 7 days of incubation. After 1 and 3 days of incubation, there were no major differences between the samples. After 7 days of incubation, important differences exist on functionalized fibres and it shows some elongated cells as a confluent layer. After 24 h, the cells are densely spread on all the samples. Cell viability and metabolic activity were investigated by Alamar Blue assay after 1, 3, and 7 days of incubation.

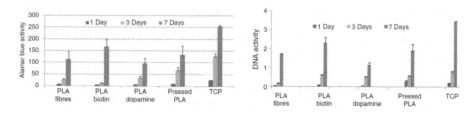

FIGURE 7.5   Metabolic activity and cell proliferation.[11]

The Alamar blue results after all the tested incubation periods show significant differences (p < 0.05) between the tissue culture plastic control and all the composite samples being used.[11]

7.5.2   FUNGAL RETTING

Fungal retting is used to extract wheat straw fibres; the fungus was taken from the bark of an elm tree. Enzymes formed by the fungus remove the pectin glue and release cellulosic fibres from the stem part.[12] In cold-water retting, anaerobic bacteria are used to break down the pectin in the plant. Straw bundles are submerged in rivers, water tanks, ponds, and vats. The process is continued for about 6–14 days, depending on the type of bacteria and temperature of the retting water. The level of environmental pollution caused by this process is high due to undesirable organic fermentation of waste waters. It produces high-quality fibres.[13] Enzymatic treatment is a very valuable and fascinating process. Enzymes can be combined with chemical and mechanical methods for different materials. Enzymes are capable catalysts and are highly specific in their work in low-energy and energy-saving circumstances.[14] To further functionalize the lignocelluloses, peroxidase-like oxidative enzymes can be used.[15] Laccase oxidize the phenolic hydroxyls compounds into phenolic radicals in the presence of oxygen.[16] The lignin content of single cellulosic fibres was decreased from 35% to 24% using the laccase process.[17] Laccase combined with natural phenols such as acetosyringone, P-coumaric acid, and syringe-aldehyde gives natural fibre–reinforced composite antimicrobial properties.[17]

7.6   CONCLUSION

Biocomposites have the ability to sense damage and act autonomously to heal it. The design optimization, mechanical performance and functional aspect expectations create complexity. Even though cost performance measures and standard testing methods will explore the way to attain a sustainable product approach. Natural fibre can be reinforced with selected engineering plastics/metals/biomaterials to develop the composites to fulfill the needs of industry requirements. Subsequently, by utilizing recycled engineering materials with right processing technique and standard quality testing procedures, ensure the functional and potential benefits. Natural fibre–reinforced composites' durability and sustainability ensured through scientific approach will meet the requirements of end users in higher end streams.

REFERENCES

1. Milne, J.M. and Reynolds, W.N. The non-destructive evaluation of composites and other materials by thermal pulse video thermography. *Proceedings SPIE* 520, 119–122, 1985.
2. Babu, B.J.C., Durai Prabhakaran, R.T. and Agrawal, V.P. Quality evaluation of resin transfer molded products. *Journal of Reinforced Plastics and Composites* 27(6), 559–581, 2008. doi: 10.1177/0731684407084211
3. Lau, S.K., Almond, D.P. and Milne, J.M. A quantitative analysis of pulse video thermography. *NDT & E International* 24(4), 195–202, 1991.
4. Tuli, S. and Mulaveesala, R. Theory of frequency modulated thermal wave imaging for nondestructive subsurface defect detection. *Applied Physics Letters* 89, 191913, 2006.
5. Banerjee, D., Chattopadhyay, S.K., Chatterjee, K., Tuli, S., Jain, N., Goyal, I. and Mukhopadhyay, S. Non-destructive testing of jute–polypropylene composite using frequency-modulated thermal wave imaging. *Journal of Thermoplastic Composite Materials* 28(4), 548–557, 2015.
6. Zakriya, G.M., Ramakrishnan, G., PalaniRajan, T. and Abinaya, D. Study of thermal properties of jute and hollow conjugated polyester fibre reinforced non-woven composite. *Journal of Industrial Textiles* 46(6), 1393–1411, 2015. doi: 10.1177/1528083715624258.2017
7. Zakriya, G.M., Ramakrishnan, G., Abinaya, D., Devi, S.B., Kumar, A.S. and Kumar, S.T. Design and development of winter over coat using Jute and hollow conjugated polyester non-woven flexible composite. *Journal of Industrial Textiles* 47(5), 781–797, 2016. doi: 10.1177/1528083716670314
8. Zakriya, G.M. and Ramakrishnan, G. Insulation and mechanical properties of jute and hollow conjugated polyester reinforced nonwoven composite. *Energy & Buildings* 158, 1544–1552, 2018. doi: 10.1016/j.enbuild.2017.11.010
9. Zakriya, G.M and Ramakrishnan G. Jute and hollow conjugated polyester composites for outdoor & indoor insulation Applications. *Journal of Natural Fibres* 16(2), 185–198, 2019. doi: 10.1080/15440478.2017.1410515
10. Sgriccia, N., Hawley, M.C. and Misra, M. Characterization of natural fiber surfaces and natural fiber composites. *Composites: Part A* 39, 1632–1637, 2008.
11. Cont, L., Grant, D., Scotchford, C., Milica, T. and Popa, C. Composite PLA scaffolds reinforced with PDO fibers for tissue engineering. *Journal of Biomaterials Applications* 27(6), 707–716, 2011.
12. Paridah, M.T., Basher, A.B., SaifulAzry, S. and Ahmed, Z. Retting process of some bast plant fibres and its effect on fibre quality: A review. *Bio Resources* 6, 5260–5281, 2011.
13. Grönqvist, S., Buchert, J., Rantanen, K. and Viikari, L. Activity of laccase on unbleached and bleached thermomechanical pulp. *Enzyme Microbial Technology* 32, 439–445, 2003.
14. Kudanga, T., Nyanhongo, G.S., Guebitz, G.M. and Burton, S. Potential applications of laccase-mediated coupling and grafting reactions: A review. *Enzyme and Microbial Technology* 48, 195–208, 2011.
15. Kim, S. and Cavaco-Paulo, A. Laccase-catalysed protein-flavonoid conjugates for flax fibre modification. *Journal of Applied Microbiology* 93, 585–600, 2012.
16. Aracri, E., Fillat, A., Colom, J.F. and Gutiérrez, A. Enzymatic grafting of simple phenols on flax and sisal pulp fibres using laccases. *Bioresource Technology* 101, 8211–8216, 2010.
17. Fillat, A., Gallardo, O., Vidal, T., Pastor, F.I., Diaz, P. and Roncero, M.B. Enzymatic grafting of natural phenols to flax fibres: Development of antimicrobial properties. *Carbohydrate Polymer* 87, 146–152, 2012.

8 Rheology and Insulation Behaviour of Composites

8.1 INTRODUCTION

Rheology is the study of the deformation behaviour of materials when a force is applied to them. Rheology is an active tool used for quality control for raw materials, manufacturing processes, and final products, and it also helps to predict material performance. The rheological properties of polymer composites are determined not only by the polymers but also by the type of fillers and their size, shape, and amount. Another key factor is the interfacial bond between fillers and polymers. Rheological properties are affected by additives and lubricants, which alter the flow properties.[1,2]

Rheometers are utilized to measure the effect of fillers on polymer composites. Rheometers are classified into two categories: rotational and capillary types. They can be parallel-plate (rotational), torque rheometer capillary, torque rheometer extension type, and melt flow indexer capillary. Four types of rheological tests can be executed utilizing the parallel-plate or rotational rheometer: (i) steady shear sweep, (ii) frequency sweep, (iii) temperature sweep, and (iv) strain sweep. The independent variables are the natural fibre used, the melt flow index of the polymer or matrix, the frequency or shear rate, the melt temperature, the strain percentage, and the gap between the plates.

Natural fibres have some disadvantages like creep (time-dependent deformation), easy absorption of moisture, flammability, thermal expansion, difficulty in application of paint, and so on. These shortcomings limit the applications of natural fibre to an extent. Additives such as coupling agents, lubricants, mineral fillers, heat or UV stabilizers, and colorants are usually utilized to overcome the problem. A coupling agent is an efficient way to enhance the compatibility between hydrophilic fibre and hydrophobic plastic.[3,4] The addition of chemical additives unavoidably changes the properties of fibre-reinforced composites. As a result, improving one property always has an effect on another property. Hence, the physical and mechanical properties of fibre-reinforced composites depend on the interactions between the fibre and polymeric materials.[5-7]

The performance of fibre-polymeric composites can be optimized by regulating the processing parameters and material preparations. A suitable combination of high-density polyethylene (HDPE), natural fibre, lubricants, and maleic anhydride-g-polyethylene fillings provides the benefits of lower shear viscosity. It aids in processing and maintaining the mechanical properties and surface smoothness of the extruded cellulosic fibre polymer composite profiles. Flow characteristics and the limited influence they exert on the melt structure according to the delivered rheological data give a good insight into the optimization of the processing variables of composite structures.[8] An isotactic polypropylene (i-PP)/wood composite in capillary rheometer shows a shear thinning phenomenon with increasing filler content.[9] Wood fibre–filled polypropylene composite rheology data show that the addition of internal lubricant reduces apparent viscosity and adding wood fibre tends to improve this phenomenon.[10] According to the rheological behaviour of a HDPE/wood fibre composite under steady shear, the extensional viscosity was not significantly affected by the wood fibre content but the rate of this phenomenon is affected by wood fibre content.[11] Composite viscosity is significantly affected by fibre length and fibre loading at lower shear rate than at higher shear rate.[12]

8.2 PHYSICAL DEFORMATION OF COMPOSITES

Composites containing a high fibre content of more than 65% weight showed various rheological properties; flax shive, HDPE and rice hull corncob, and walnut shell flour show a high complex viscosity, loss and storage moduli with less damping factors.[13] Composites made up of PLA and LDPE contain lower amounts of cellulose exhibiting good storage and loss moduli compared to other matrix polymers.[14] Bond strength can be effectively enhanced by altering the fibre surface through physical or chemical surface treatments such as corona treatment, plasma treatment, silanization, heat treatment, and acetylation. An unchanging torque is also an indicator of the homogenization of the filler in the melt of the composite structure.[15]

Torque rheometry curves of agro fibre/HDPE composites are shown in Figure 8.1. Weights of 65%, 32%, and 3% of 50 g of Corncob fibre (CCF), rice hull fibre (RHF), walnut shell fibre (WSF), and flax shive fibre (FSF) with HDPE and a lubricant were mixed together to formed composites. The resulting rheometry curves are clearly depicted in Figure 8.1. The torque rheometer provides quantitative data that 65% weight of filler allows for the maximum torque at steady state. The torque decreases with a decrease in agro fibre sizes. IT was observed that within 2–4 minutes of mixing time, flax shive and rice hull fillers were completely dispersed in the polymer matrix, but corncob and walnut shell fillers took 4–6 minutes to disperse. Thus, the same filler load of corncob and walnut shell fibres showed a higher torque value than flax and rice hull fibers.[13,16,17] In a strain sweep test, the storage modulus G′ changed within the entire strain above 0.02%. The decrease in the storage modulus with amplified strain indicates breakage of the composite material structure. This is due to poor compatibility and ineffective bonding between the matrix and fibre fillers.[13,18]

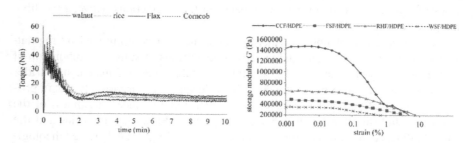

FIGURE 8.1 Torque rheometry curves of agro-fiber/HDPE composites and its storage modulus versus strain.[13]

8.2.1 Complex Viscosity of Fibre Composite

The addition of filler to the polymer matrix improves the viscosity of the melt. Rigid fibre particles disturb the run flow of the polymer melt. The improved viscosity depends upon the particle size distribution, concentration, and shape of the fillers. A steady decrease in complex viscosity with increased shear rate or frequency means that the agro fibres are exhibiting shear thinning behaviour like pseudoplasticity.[13] A reduction in particle size improves viscosity; vice-versa, an increase in particle size reduces viscosity. Particles with a broader size distribution or span will have a higher amount of free space to wander, and the sample flow becomes easier. Varying the span, changing the number of particle–particle interactions, and changing the lubricants considerably improve the output. Widening the process window and lowering the melt temperature are techniques that are widely used to prevent polyolefin's melt fracture, sharkskin, and flow instability.[13,19]

8.2.2 Storage modulus of fibre composites

A high fibre load of 65% fibre in the bulk of the melt causes discontinuity increases. Different polarity between fibres and HDPE creates migration of polymer from the surface and its continuity increases to upkeep the melt exit from the die. This steadiness enables elastic energy recovery and improves the melt elasticity. The storage modulus performance indicates the ability to store the energy of external forces in the fibres. Corncob, rice hull, and flax shive composites show higher storage modulus. The storage modulus of walnut shell composites decreases with an increase in angular frequency (ω). The more smaller particles that result, the more particle–particle interactions improve resistance to flow.[20]

8.2.3 Loss Modulus of Fibre Composites

Corncob composites show a high impact-absorption quality compared to rice hull and flax shive composites. The dissimilarities in loss modulus between the fibres are due to differences in their particle size and distribution. As frequency increases, the loss modulus differences increase between composites, and the same happens in low frequency ranges. The loss modulus behaviour indicates that impact absorption improves for all samples.[13,21,22]

8.2.4 Damping Factor of Composites (tan ∂)

The differences in tan ∂ arises for the composites due to the particle size, particle size distribution, and bulk density of fibre fillers. Tan ∂ is the relation between G''/G' and with incorporation of fillers along with polymers G' increased and G'' decreased due to the fact that energy loss is reduced by porous fillers, hence the value of tan ∂ decreased. The flattened section in the curve indicates the relaxation state of the particles.[13,23,24] All four kinds of deformed rheological curve of composites are shown in Figure 8.2.

A resin matrix composite of polybutylene terephthalate (PBT) fibres and polylactic acid (PLA) is prepared using a melt-stretching and rapid cooling method.

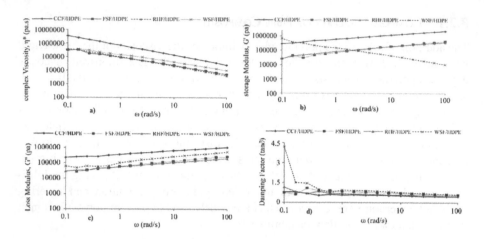

FIGURE 8.2   a) Complex viscosity with frequency, b) storage modulus with frequency, c) Loss modulus with frequency, d) damping factor with frequency.[13]

The melting point of PBT is higher than that of PLA using this phenomenon, and the rheological properties of the composites are evaluated between the melting points of PLA and PBT. A weight of 1% of PBT fibres significantly enhances the elongational viscosity, while shear viscosity is hardly changed. Elastic deformation of the network of flexible PBT fibres leads to bending of fibres and/or friction between fibres, and is responsible for strain-hardening. The surplus shear deformation of the matrix liquid between fibres also impacts the high level of elongational viscosity to a certain degree.[25]

8.3   INSULATION BEHAVIOR OF COMPOSITES

An acoustic rheometer employs a piezo-electric crystal. It can easily launch a continuous wave of extensions and contractions into the fluids. Working on a noncontact basis, it relates an oscillating extensional stress. Acoustic rheometers are used to measure the sound speed and attenuation of ultrasound for a set of frequencies in the megahertz range. Sound speed is also considered as a measure of system elasticity. It can be converted into fluid compressibility. Attenuation is a measure of viscous properties. It can be converted into viscous longitudinal modulus. In the case of a Newtonian liquid, attenuation yields information on the volume viscosity. This type of rheometer works at higher frequencies compared to other kinds. It is more appropriate than any other rheometer for studying effects with much shorter relaxation times.[26]

Composites with sound absorption properties were prepared using a pressing process of fibrous multilayer structures. Three types of composites were prepared and compared: composites reinforced by cotton fibres, composites reinforced by cellulose ultra-short/ultra-fine fibres, and composites reinforced by cotton fibres/cellulose ultra-short/ultra-fine fibres. The roles of the cellulose ultra-short and ultra-fine fibres in altering the sound absorption properties of the composites were analyzed. The use of natural fibres with a thermoplastic polymer results in augmented sound absorption. The best quality of sound absorption is obtained by blending cotton fibres/cellulose ultra-short/ultra-fine fibres, especially nanofibers, as a reinforcement through needle punching technology.[27]

An increase in thickness and a reduction in air permeability enhances the sound-absorption properties of the composite material. A blend of 70% cotton fibre and 30% polyester fibre achieves the best sound absorption coefficient in mid- to high-frequency ranges. The higher the number of fibres per unit area, the better the sound absorption quality of the composite material. A blend of acrylic/polypropylene and cotton/polyester fibre improves the sound-absorption quality of the composite material in the low- to mid-frequency ranges.[28] Figure 8.3 shows the sound-absorption coefficients of different kinds of composites and their different manufacturing methods with respect to frequency in Hz.

The effectiveness of blending recycled natural fibres blended with synthetic fibres has been confirmed acoustically. Biocomposites from agricultural wastes, such as rice straw and sawdust, have also been examined. Recycled or reused natural fibres blended with artificial nonwoven fibres demonstrate high sound absorption coefficients at high-frequency ranges from 2000–6300 Hz, low sound-absorption coefficients at low-frequency ranges from 100–400 Hz, and better sound absorption coefficients at mid-frequency ranges from 500–1600 Hz. The sound-absorption coefficient of composites or nonwovens at all frequency ranges was improved by improving the thickness factor. Adding air traps or spaces within the composite shows improved sound absorption at low, mid frequencies ranges and little more at high sound resulted in absorption in high-frequency ranges. Recycled fibrous materials have the potential to be low-cost, lightweight, and biodegradable sound-absorption materials.[29]

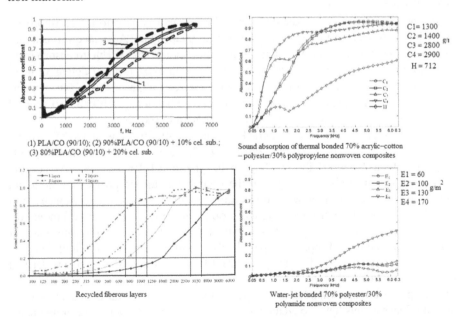

(1) PLA/CO (90/10); (2) 90%PLA/CO (90/10) + 10% cel. sub.; (3) 80%PLA/CO (90/10) + 20% cel. sub.

Sound absorption of thermal bonded 70% acrylic–cotton – polyester/30% polypropylene nonwoven composites

Recycled fiberous layers

Water-jet bonded 70% polyester/30% polyamide nonwoven composites

FIGURE 8.3 Sound absorption coefficients of different kinds of composites.[27–29]

8.4 SURFACE COATING OF COMPOSITES

Natural fibre–reinforced polymeric composites require protective coatings and barrier layers. Coatings or surface finishes will be a precautionary solution to the long-term durability and appearance of fibre-reinforced composite. There is a wide range of surface finishes, like gel coatings, adhesives, and surface veils. Other protective finishes include aliphatic isocyanates, polyesters, acrylics, polyurethanes, and epoxies, and in some cases fine sand is added for supplementary protection. A designed bulkiness of composites has appropriate surface characteristics for clinical and cleaning applications. A biocompatible modified layer with suitable wear and corrosion resistance would mitigate problems. The mechanical properties of biomaterials are dictated by their bulk properties, and tissue biomaterial interactions are governed by surface properties. Less than 1 nm surface coatings protect the bulk phenomenon of composites.[30] Air release requirements, thick film build-up, colour technology, and rapid cure times allow the production of in-mould finished composite surfaces with integrity, retaining excellent gloss and colour after years of environmental exposure.[31]

A common surface coating or finish for fibre-reinforced composites is a gel coat. This is specifically formulated with polyester resin and is generally applied to the composite material to improve chemical resistance, improve abrasion resistance, and provide a moisture barrier. Gel coats are used to enhance the look of products such as the surface of a boat hull or golf cart. A unique benefit of gel coats is that they are available in many colours, created by the incorporation of pigments as per the specification of an engineer. They protect the look of the product and provides weather resistance for outdoor products. They filter out ultraviolet (UV) radiation, improve chemical resistance, add flame resistance, improve abrasion resistance, and can be layered as a moisture barrier.[32,33]

In fibre reinforced composite designs, a surface veil is used to afford improved corrosion resistance or a weatherproof barrier to the product. A surface veil is a fabric made from nylon or polyester fibre. It is a thin sponge and is able to absorb resin to up to 90% of its volume. It also helps to retain an extra layer of defensive resin on the fibre-reinforced composite surface. Surface veils improve the surface appearance and ensure the presence of corrosion resistance barriers. Veils can also be used concurrently with gel coats to provide reinforcement to the resin.[32,33]

Peeling, potential chalking, and cracking that eventually leads to loss of resin on the composite surface is called *fibre blooming*. The rate of UV degradation depends on geographical location, fibre loading, resin type, and filler packages. An acrylic, epoxy, or urethanes binder coating on the composite surface provides a moderate temperature resistance along with good strength, a high temperature resistance along with higher strength and good toughness obtained through respective binder formulations. A surface coating of 10–20 mm depth protects the composite material from UV degradation. Properly prepared Fibre reinforced polymer (FRP) composites accept a wide variety of surface painting like oil and water-based paints derived from acrylic and urethanes. Paints need not be breathable and no extraordinary surface preparation is mandatory. Proper abrasion and removal of residuals from mould-release agents helps to create an even surface finish for adhesive or mechanical attachment.[32]

It is expected that technical developments in biofunctionalization, micro- and nanoengineering, synthetic chemistry, and surface characterization will lead to the development of advanced and bioactive composite surfaces. In particular, surfaces that are highly topographically complex at the submicron level conveniently interact with some proteins effectively. This will provide unparalleled control over cellular activities and healing responses. A surface modification process is overly complicated; it is more suitable to minimize the number of steps or to design each step to be relatively insensitive to small changes in the experimental conditions.[33]

8.5 CONCLUSION

Polylactic acid (PLA), polyhydroxyl alkanoates (PHA), polyhydroxylbutyrate (PHB) and its copolymers poly(hydroxylbutyrate-co-valerate) (PHBV), cellulose, and starch-derived plastics are produced from renewable agricultural and biomass feedstock. Among them, PLA is the biodegradable polymer that has comparable strength and stiffness relative to petroleum-based polymers.[34] Since significant quantities of organic waste from industry and agriculture are unutilized, utilization of organic residue materials in bio- and natural composites through eco-friendly manufacturing methods can reduce the cost of the products.[35] Fibre type, particle size, content, and shape affect the rheological qualities of the final composites.[2] PLA can be processed with potato pulp powder and a plasticizer of acetyl tributyl citrate (ATBC). Calcium carbonate ($CaCO_3$) can be used in low percentages as an inert filler to simplify the removal of injection-moulded specimens from the mould. The rheological properties of PLA-based matrices depend on the type of reinforcing fibres used.[36]

REFERENCES

1. Mezger, T.G. *Rheology Handbook.* (2nd ed.). Hanover, Germany: Vincentz Network, 2006.
2. Ogah, O.A. Rheological properties of natural fiber polymer composites. *MOJ Polymer Science* 1(4), 147–148, 2017. doi: 10.15406/mojps.2017.01.00022
3. Wang, Y., Cao, J.Z. and Zhu, L.Z. Interfacial compatibility of wood flour/ polypropylene composites by stress relaxation method. *Journal of Applied Polymer Science* 126(S1), E89–E95, 2012.
4. Rodriguez Llamazares, S., Zuniga, A. and Castano, J. Comparative study of maleated polypropylene as a coupling agent for recycled low density polyethylene/wood flour composites. *Journal of Applied Polymer Science* 122(3), 1731–1744, 2011.
5. Arino, R. and Boldizar, A. Processing and mechanical properties of thermoplastic composites based on cellulose fibers and ethylene-acrylic acid copolymer. *Journal of Polymer Engineering Science* 52(9), 1951–1957, 2012.
6. Petchwattana, N. and Covavisaruch, S. Influences of modified chemical agents on foaming wood plastic composites prepared from poly (vinyl chloride) and rice hull. *Journal Material Design* 32(306–307), 869–873, 2011.
7. Leu, S.Y., Yang, T.H. and Lo, S.F. Optimized material composition to improve the physical and mechanical properties of extruded wood-plastic composites. *Journal of Construction and Building Materials* 29, 120–127, 2012.
8. Adhikary, K.B., Park, C.B. and Islam, M.R. Effects of lubricant content on extrusion processing and mechanical properties of wood flour high density composites. *Journal of Thermoplastic Composites* 24(2), 155–171, 2011.

9. Maiti, S.N., Subbarao, R. and Ibrahim, M.D. Effect of wood fibers on the rheological properties of i-pp/wood fiber composites. *Journal of Applied Polymer Science* 91(1), 644–650, 2004.

10. Harper, D. and Wolcott, M. Interaction between coupling agents and lubricants in wood-polypropylene composites. *Composites Part A* 35, 385–394, 2004.

11. Li, T.Q. and Wolcott, M.P. Rheology of wood plastics melt part 1 capillary Rheometry of HDPE filled with maple. *Journal of Polymer Engineering & Science* 45(4), 549–559, 2005.

12. George, J., Janardhan, R., Anand, J.S., Bhagawan, S.S., and Thomas, S. Melt rheological behaviour of short pineapple fibre reinforced low density polyethylene composites. *Polymer* 37(24), 5421–5431, 1996.

13. Ogah, A.O., Afiukwa, J.N. and Nduji, A.A. Characterization and comparison of rheological properties of agro fiber filled high density polyethylene biocomposites. *Open Journal of Polymer Chemistry* 4(1), 12–19, 2014.

14. Shumigin, D., Tarasova, E., Krumme, A. and Meier P. Rheological and mechanical properties of poly (lactic acid)/cellulose and LDPE/cellulose composites. *Journal of Materials Science* 17(1), 32–37, 2011.

15. Joseph, P.V., Joseph, K. and Thomas, S. Effect of processing variables on the mechanical properties of sisal fibre reinforced polypropylene composites. *Composite Science and Technology* 59, 1625, 1999.

16. Liang, J.Z. Effects of extrusion conditions on melt viscoelasticity during capillary flow of low-density polyethylene. *Journal of Thermoplastic Composite Materials* 22(1), 99–110, 2000.

17. Othman, N., Ismail, H. and Mariatti, M. Effect of compatibilizers on mechanical and thermal properties of bentonite filled polypropylene composites. *Polymer Degradation and Stability* 91(2), 1761–1774, 2006.

18. Aranguren, M.J., Mora, E., DeGroot, J.V. and Macosko, C.W. Effect of reinforcing fillers on the rheology of polymer melts. *Journal of Rheology* 36(6), 1165–1182, 1992.

19. Mohanty, S. and Nayak, S.K. Mechanical and rheological characterization of treated jute-HDPE composites with a different morphology. *Journal of Reinforced Plastics and Composites* 25(13), 1419–1439, 2006.

20. Son, J., Gardner, D.J., O'Neill, S. and Metaxas, C. Understanding the viscoelastic properties of extruded polypropylene wood plastic composites. *Journal of Applied Polymer Science* 89(6), 1638–1644, 2003.

21. Nair, K.C.M., Kumar, P.P., Thomas, S., Schit, S.C. and Ramamurthy, K. Rheological behavior of short sisal fiber-reinforced polystyrene composites. *Composites Part A: Applied Science and Manufacturing* 31(11), 1231–1242, 2000.

22. Mezger, T.G. *Rheology Handbook*. (2nd ed.). Hanover, Germany: Vincentz Network, 2006, pp. 41,42.

23. Mir, S., Yasin, T., Halley, T.J., Siddiqi, H.M. and Nicholson, T. Thermal, rheological, mechanical and morphological behavior of HDPE/chitosan blend. *Carbohydrate Polymers* 83(2), 414–421, 2011.

24. Pan, M.Z., Zhang, S.Y. and Zhou, D.G. Preparation and properties of wheat straw fiber-polypropylene composites. Part II. Investigation of surface treatments on the thermomechanical and rheological properties of the composites. *Journal of Composite Materials* 44(9), 1061–1074, 2010.

25. Yokohara, T., Nobukawa, S. and Yamaguchi, M. Rheological properties of polymer composites with flexible fine fibers. *Journal of Rheology* 55(6), 1205–1218, 2011.

26. https://en.wikipedia.org/wiki/Rheometer

27. Krucińska, I., Gliścińska, E., Michalak, M., Ciechańska, D., Kazimierczak, J. and Bloda, A. Sound-absorbing green composites based on cellulose ultra-short/ultra-fine fibers. *Textile Research Journal* 85(6), 646–657, 2015.

28. Küçük, M. and Korkmaz, Y. The effect of physical parameters on sound absorption properties of natural fiber mixed nonwoven composites. *Textile Research Journal* 82(20), 2043–2053, 2012.
29. Seddeq, H.S., Aly, N.M., Marwa, A.A. and Elshakankery, M.H. Investigation on sound absorption properties for recycled fibrous materials. *Journal of Industrial Textiles* 43(1), 56–73, 2012.
30. Ong, J.L. and Lucas, L.C. Auger electron spectroscopy and its use for the characterization of titanium and hydroxyapatite surfaces. *Biomaterials* 19, 455–464, 1998.
31. Published courtesy of Dr L S Norwood, Scott Bader Company Ltd., Coatings., access on Thursday, 24th January 2019. Available from: https://netcomposites.com/guide/coatings/
32. http://compositeslab.com/composite-materials/surface-finishes/ (accessed: 23/03/2020).
33. Nouri, A. and Wen, C. *Introduction to Surface Coating and Modification for Metallic Biomaterials*. Cambridge, UK: Elsevier Ltd., 2015.
34. Hamad, K., Kaseem, M., Ayyoob, M., Joo, J. and Deri, F. Polylactic acid blends: The future of green, light and tough. *Progress in Polymer Science* 85, 83–127, 2018.
35. Väisänen, T., Haapala, A., Lappalainen, R. and Tomppo, L. Utilization of agricultural and forest industry waste and residues in natural fiber-polymer composites: A review. *Waste Management* 54, 62–73, 2016.
36. Mekonnen, T., Mussone, P., Khalil, H. and Bressler, D. Progress in bio-based plastics and plasticizing modifications. *Journal of Materials Chemistry A* 1, 13379–13398, 2013.

9 Application of Composites in Engineering

9.1 INTRODUCTION

Natural fibre–reinforced composites are currently used in several engineering domains.[1] This is undoubtedly because of the biodegradable nature of natural fibres, the recyclability of polymers, the low weight of the materials, ease of processing, and cost reductions.[2] High specific properties of natural fibre–reinforced polymer composites include low thermal conductivity, good strength-to-weight ratio, high impact strength, nonconductivity, nonmagnetivity, corrosion resistance, dimensional stability, design flexibility, part consolidation, and radar-transparency.[3] Fibre-reinforced composites are extensively required where the need for mechanical strength is one of the significant parameters. Industries such as automotive, aerospace, marine, and oil and gas extensively employ fibre-reinforced composite materials, and such industries are considered to be key players in the global market for composites. Composites are also utilized in the electrical and electronics industries. The composites market for 2016 was valued at USD 76 billion, and this figure is expected to rise on the basis of a compound annual growth rate (CAGR) of 8.9% between 2017 and 2025.[4] The automotive polymer composite market for 2016 was valued at USD 6.4 billion, and this figure is expected to rise based on a compound annual growth rate (CAGR) of 6.8% to an estimated value of USD 11.62 billion by 2025.[4] The future market performance of fibre-reinforced composites is predicted in sectors such as defence, aerospace, construction, wind-power plant, and automobile and transportation. This can help to reduce weight in automobiles, enhancing fuel efficiency and reducing hazardous emissions. Fibre-reinforced composites could reduce weight in the range of 15% to 40% depending on the kind of reinforcement and the technique used.[5]

Transport, conveyance, and logistics are hugely reliant on petroleum-based fuels. A statement released by Organization of the Petroleum Exporting Countries (OPEC) secretariat said that in 2017, worldwide demand for oil-based transportation was 6.3 mb/d (million barrel per day) for air transportation, 4 mb/d for maritime transportation, 1.8 mb/d for rail transportation, and 43.6 mb/d for road transportation. The number of passenger vehicles is anticipated to rise from 1102 million in 2017 to 1980 million in 2040, while the number of commercial vehicles may rise from 230 million to 462 million.[6] The total transport segment is accountable for 23% of global CO_2 emissions; of this, road transport contributes 72%, air transport 11%, and maritime transport 11% of emissions.[7] Fuel intake and CO_2 emissions can be reduced by reducing automobile weight. A reduction of 10% in automobile weight improves fuel efficiency by approximately 7%, and a minimum 1 kg reduction in automobile weight lessens the CO_2 emissions released into the environment by up to 20 kg.[8–10] Accordingly, motor-vehicle manufacturers are making efforts to reduce vehicle weight by changing the materials they use. However, there are rigid rules and regulations in the European Union and in Asian countries concerning the ecological impact of the automotive lifecycle and end-of-life requirements, including raw material selection, manufacturing-cum-processing, and disposal.[11–13] Automobile companies have already shifted their manufacturing capabilities from steel alloys to aluminium alloys, and they are now moving on from the potential of aluminium towards the use of fibre-reinforced polymer composites for certain applications.[14]

Fibre-reinforced composites have several kinds of construction features with a wide range of variables, such as the appropriate selection of material components, fibre density, proper fibre orientation, composition of composite materials, composite thickness, composite density, composite structure, fabrication methods, and the use of mixed or hybrid reinforcements from natural or synthetic sources.[15] In spite of their many advantages, natural fibres have some limitations such as high flammability, high moisture absorption, low strength, and low modulus. By using a hybridization technique, the drawbacks of natural fibre composite could be negated and a new class of material developed with desirable advantages. Hybrid composites retain the benefits of constituent materials while improving their lifetime, load-bearing ability, and electrical, thermal, mechanical, and structural properties with reduced cost.[15]

9.2 AIRCRAFT

Natural fibre–reinforced polymer composites offer good aircraft efficiency through lesser fuel consumption and reduced emissions. The high strength and stiffness of natural fibres allow the fabrication of complex shapes, enabling improvements in aerodynamic efficiency. The good corrosion and fatigue-resistant properties of natural fibres have led to their use to produce more than 200 aircraft components, including propeller systems and side walls.[16,17] Flax fibre and recycled thermoplastic sheets are used to produce aviation parts by Invent GmbH, Aimplas, and Lineo in a collaboration with an EU co-funded project along with Boeing Research and Technology.[18] Another EU-funded project called Eco-Compass is carrying out research on producing eco-friendly composites from bio-sources and recycled materials for the aviation sector. Flax and ramie fibres have shown potential for use with bio-resins/recycled carbon fibres for aircraft secondary structures and interior applications.[19]

Epoxy resin is generally used for high-performance light-weight application purposes in combination with carbon fibre/natural fibres. In the Boeing 737, sidewall panels were manufactured with flax/epoxy sandwich composites. Flax fabrics were additionally treated with halogen-free fire retardants. The flax/epoxy prepregs used in this process are 35% lighter than carbon/epoxy prepregs and have a similar cost to glass/epoxy unidirectional prepregs.[20]

Engineering thermoplastics such as polyether ether ketone, polyphenylene sulphide, Polyamide (PA), and Polycarbonate (PC) are mainly used in the aviation industry, due to their good fire- and smoke-resistant properties, complying with the flame and toxicity regulations of the aviation sector. For example, Victrex Europa

GmbH developed VICTREX® PEEK (polyether ether ketone) reinforced with chopped glass/carbon fibre for aircraft applications. It has improved mechanical strength and dimensional stability comparable to metal alloys. It can withstand high temperatures and has a melting point of 343°C. In addition, the material is corrosion resistant, chemical resistant, wear resistant, and abrasion resistant.[18,20,21]

9.3 MARINE

The introduction of fibre-reinforced polymer (FRP) composites in the maritime industry was driven by the need to find a substitute for wood. Wood degrades and is subject to attack and degradation by biological agents. Steel and aluminium alloys can corrode and create problems during the welding process. To further reduce the weight of maritime vehicle construction, fibre-reinforced composites material play a significant role in maritime applications. FRP composites are used in both internal and external equipment such as bulkheads, ducts, decks, heat exchangers, propellers, valves, pumps, watertight doors, and pipes. Composites of glass/carbon-reinforced polymers are utilized in racing boats and ships. These materials offer lightweight benefits in construction and provide good wave impact resistance.[20]

A reduction in weight helps to reduce fuel consumption and improves efficiency. FRP composites for maritime vessels need to be water and corrosion resistant. Carbon fibre–based composites absorb electromagnetic waves and therefore gives stealth characteristics in the making of boat hulls. Epoxy, polyester, vinyl ester and phenolic based polymer resins reinforced with carbon fibres are generally used to produce racing boats and yachts.[21,22] The main disadvantages of carbon fibre–reinforced polymer composites in maritime applications are environmental concerns, cost, the issue of recyclability, and poor reparability. Commonly used thermosetting polymer resins such as polyester, epoxy, and vinyl ester are difficult to recycle. Currently, recyclable thermoplastics such as PA, PP, PET, and PBT are preferably used in maritime construction.[23]

Waste disposal and environmental issues have created interest in the use of biocomposites to make boats. Flax, cotton, and hemp fibres are commonly used in the weaving of canvas sails, nets, and ropes. Flax fibre has better vibration absorption properties than glass or Kevlar fibre. This quality makes flax fibre reasonably good for vessels, which are subjected to high vibration levels when sailing at high speed. The moisture absorption properties of natural fibre composite are a major concern, but this can be overcome by applying a waterproof coating of paraffin wax or linseed oil.[24]

The maritime industry still dominated by the use of carbon and glass fibre, and so the Nav Eco Mat joint research project was initiated by marine construction companies and material research laboratories to investigate the high usage of natural fibres in future maritime applications. An eco-friendly composite has been developed using PLA and flax fibres, and it shows similar properties to glass or carbon composite material. Naskapi-style canoes have also been made of composites comprising PLA films reinforced with flax fibre mats. Expression of people to the environment is summarized and is given in Table 9.1 for the consideration eco processing. In other research, 50% flax fibre reinforced with an epoxy matrix was used to produce

TABLE 9.1

Expression of people to environment summary[5]

Measures	Description
Eco-friendly	The component is safe to use, generating fewer emissions both during preparation and when in use.
Free from hazardous substances	Materials or parts of the component do not contain toxic or hazardous substances.
Reliability	The component consists of reliable materials that have higher fatigue strength to ensure the lifetime of the component.
Long lifetime	The component must be able to encounter multiple loading including bending and torsional.
Price	The lower price is desirable.
Durability	Materials or some parts of the component have the robustness to encounter multiple loading.
Lightweight	Materials and some parts of the component are from non-heavy material.
Easy to maintain	The component is easy for the user to handle and maintain.
Impact resistance	The component is able to secure the position of the vehicle during a crash or high-impact loading on the vehicle.
Easy to reuse	Materials or some parts of the component could be reused for another purpose
Easy to recycle	Materials or some parts of the component could be recycled and used for another purpose.
Easy to manufacture	Materials of the components are ready either in raw or processed condition and some parts of the component are machinable.
Not easy to break	The component keeps its shape if the vehicle rolls over and maintains its connection with other linked components.
Less transportation	Materials or some parts of the component are available within a reasonable radius and require less transportation during the manufacturing process.
Fewer materials	Design of the component must reduce the number of different types of material used.

'Araldite', a racing-boat prototype, while composite of flax fibres with bio-resin, under the brand name EcoComp® UV-L, is used to manufacture kayaks and canoes. Basalt fibres, which have good mechanical characteristics and are eco-friendly, are now being considered for maritime applications. Basalt and balsa-wood composites have been used to produce decks for sailing and hull yachts.[20,24]

9.4 AUTOMOBILE

Materials should be amended to improve the crashworthiness of automobiles by absorbing impact energy at the time of collision and ensuring passenger safety. Every year numerous vehicles reach the end of their useful life, generating 8–9 million tons of waste, of which only 65%–75% is recycled successfully. The European Union has provided a guideline for the end-of-life of automobile vehicles. It recommends that vehicles be composed of 95% recyclable material, with 85% recoverable by mechanical recycling and 10% through energy recovery/thermal recycling methods.[9,25–27] Natural fibre–reinforced polymer composite can be used in internal and external components of automobile vehicles, such as door linings, panels for floor and doors, luggage compartments, headliners of vehicle roofs, backrests of seats, and parcel shelves. Kenaf, jute, and ramie-reinforced composites are mostly used in India and Asia. Flax and hemp reinforced composites are used in Europe. Ramie and sisal reinforced composites are used by South American automobile producers. Natural fibre composites reduce automobile weight and improve efficiency without compromising the safety features of the vehicle, saving approximately 30% of the vehicle's total weight and 20% of the cost.[28]

For decades, European car fabricators have been using natural fibres such as sisal, jute, flax, and hemp to produce dashboards, door panels, and headliners, resulting in lower CO_2 emissions. Nylon/flax composites are used to make floor mats and hemp/cotton composites are used to make seat-back lining, interior cladding, and floor panels. Flax composites are used as an alternative to asbestos in the production of car disk brakes. Coconut fibres are used to make back cushions, seat bottoms, and head restraints. Abaca fibres are used to make floor body panels and kenaf is used to make door inner panels.[27] The typical amounts of natural fibres used in the automobile industry are shown in Table 9.2.

German car manufacturers like Audi, BMW, Daimler Chrysler, Mercedes, and Volkswagen use natural fibre–reinforced composites in their vehicles. International car manufacturers like Ford, GM, Opel, Peugeot, Renault, and Volvo are following the trend of utilizing natural fibre composites in their automobiles. The Travego coach from Mercedes-Benz uses flax/PP-reinforced composite in its engine encapsulation panels. The S-Class Mercedes-Benz uses 35% polyurethane (PU) and a 65% blend of flax/hemp/sisal fibre composite for making inner door panels. BMW is using biocomposites in structural components such as fender liners, bumpers, and suspension parts. Acrylic reinforced with sisal fibre mats are used in the inner door panels of its 7 series sedan.[21,30–33] The use of natural fibres in automotive components is clearly depicted in Table 9.3.

Toyota mostly uses kenaf fibres reinforced with a polylactic acid (PLA) matrix for its interior components and it has introduced a new bio-PET material made from sugarcane for the manufacture of parcel shelves and luggage trunk liners.

TABLE 9.2
Amount of natural fibres used in automobile industry[29]

Parts of automobile	Amount of natural fibres
Front door linens	1.2–1.8 kg
Rear door linens	0.8–1.5 kg
Boot linens	1.5–2.5 kg
Parcel shelves/packaging tray	Up to 2 kg
Seat backs	1.6–2.0 kg
Sunroof sliders	Up to 0.4 kg
Headliners	2.5 kg

TABLE 9.3

Natural fibres in automotive components[8,28–30]

Manufacturers	Applications	Natural fibre
BMW	Door panels, boot linings, seat backs, headliner panels, noise insulation panels	Flax, sisal, cotton, wood, hemp
Audi	Seat backs, back door panels, side door panels, boot liners, spare tire liners	Flax, sisal
Volkswagen	Seat backs, door panels, boot liners, boot lid finish panels	Flax, sisal
Ford	Door panels, boot liners, floor trays, door inserts	Kenaf, wheat, castor
Toyota	Floor mats, spare tire covers, door panels and seat backs, luggage compartments	Kenaf, sugarcane, Bamboo
General Motors	Seat backs, cargo area floor mats, noise insulation, door panels, trim	Cotton, flax, wood, kenaf, hemp
Opel	Door panels, head liner panels, instrumental panels	Flax, kenaf
Lotus	Body panels, interior mats, seats	Hemp, sisal
Daimler Chrysler	Door panels, floor panels, trunk panels, dashboards, pillar cover panels, seat back rests, insulation	Flax, sisal, coir, wood, banana, cotton

Using a hemp/flax/kenaf/sisal mixture, the Findlay industry produces headliners for the Mack truck. Interior components of Mitsubishi motor vehicles are produced from bamboo fibre–reinforced polybutylene succinate bio-resin, and in collaboration with Fiat SPA, the firm has developed floor mats from nylon fibres/bio-PLA produced from sugarcane molasses/PP fibre–reinforced composites. High strength PA is used to develop thin-walled front-end carriers for the Skoda Octavia with good surface quality and high stiffness. Lotus has introduced a model car named Eco Elise, which uses hemp fibre reinforced with polyester composite for body panels, woven sisal fibres in carpet, and eco-wool in upholstery interiors.[8,21,28–33] A photograph in Figure 9.1 shows the contribution of natural fibre composites in the automobile sector.

9.5   BUILDING CONSTRUCTION

The method of using natural fibres to increase or enhance the strength possessions of soil blocks is extensively used in the field of construction and building materials. Natural fibre soil composites have the benefits of being able to produce low-embodied-energy houses with better occupant comfort and minimal impact on the environment. Fibre-reinforced soil composite comprises soil mass that contains randomly distributed discrete fibre elements. It provides an enhancement in the mechanical enactment of the fibre and soil composite.[35] The enactment of the fibre-reinforced soil matrix relies on factors such as fibre type, aspect ratio of fibres, fibre characteristics, particle size, distribution of soil, and proportion of fibre to soil. Natural fibres can be used as reinforcement to enhance the engineering properties of different types of soil for use in building eco-friendly structures.[36,37]

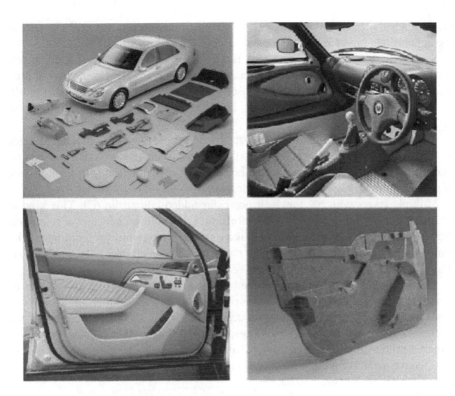

FIGURE 9.1   Natural fibre reinforced composites in automobile sectors.[11,34]

Coconut fibre–reinforced soil blocks performed better in terms of compressive strength than oil palm and bagasse fibre–reinforced soil blocks. The pull out for all kinds of natural fibre lengths can be clarified by the fibre's poor interfacial bond with the soil matrix and in some cases the shortness of the natural fibres inserted in the soil matrix. This can result in pull-out or rupture of the natural fibres.[37–39] Jute and Kenaf natural fibre–reinforced composites have been utilized globally in interior design applications such as the manufacture of furniture and home panelling.[16,40] There are clearly opportunities for applications of plant fibre products in building and construction materials. Further research into natural fibre's miscellaneous characteristics, the use of bio polymers and resin, and meeting the different demands for building construction applications is essential. The waste management of natural plant fibre composite is expected to have strategic benefits and it would be of great use to support building and construction.[41,42]

The decay temperature of natural fibres is very low at 240°C. However, the components of fibre, like cellulose, hemicellulose, and lignin, degrade at different temperature ranges – hemicellulose decomposes between 200°C and 260°C and cellulose between 260°C and 350°C, and lignin slowly decomposes from 200°C to 500°C. The decomposition behaviour of jute and sisal fibre is comparatively good for use in the building and construction sectors. Jute/HCP composites were made into four kinds of structure and the effects of natural weathering, thermal, sound insulation and mechanical properties on the structures were analyzsed.[43] The following conclusions can be made from the results of the experiments:

Sandwich structure composite (A): Before natural weathering, this composite shows good thermal conductivity and noise reduction coefficient compared to after weathering. This composite could be used as an alternative for interior trims like wall panels and ceilings in auditoriums, music venues, and cinema theatres, and so on.

Blended nonwoven structure composite (B): This kind of composite is not affected by the natural weathering process. It could be used to make side wall coverings for interior building blocks, furniture, and household racks and, it may be used instead of wood as a base material for interior design.

Multiple layers with 5% low-melt polyester added composite (C): This kind of composite shows better post-impact performance and an average value of sound and thermal insulation. It may be used to make outer wall coverings for building blocks, window frames, and door panels and to make articles which may reduce the utilization of plastics, metals, and alloys.

Multiple layers of nonwoven stitched composite (D): Exposure to weathering causes this kind of composite to deteriorate easily. Before weathering, it shows its highest thermal insulation and noise reduction coefficient. It is recommended for use in the interiors of commercial buildings, air-conditioned rooms, sound recording rooms, seminar hall, and so on, which will help to reduce energy costs and will be an effective means of energy conservation.[43]

9.6 GENERAL APPLICATIONS

Jute-reinforced polymer composite can be used for producing windows, doors, corrugated sheets, other types of furniture, and water pipes. Kenaf-reinforced composites can be used to make packing shelves for the automobile sector and any other kind of article. Due to its good strength, durability, high tenacity, and resistance to salt water, sisal fibre–reinforced composites can be used in cement matrix. Abaca fibres are used in the production of floor panels as they possess excellent tensile strength. Coir fibre–reinforced composites have a range of properties such as resistance to fungi and pests, high wear resistance, and good acoustic resistance; they are used to produce helmets, roofs, postboxes, insulation panels, and bullet-proof vests. Rice husk and straw fibre–reinforced composites are used to produce structural board applications. General wood fibre composite is used to produce sliding door inserts and cargo floor panels.[44]

9.7 CONCLUSION

Natural fibre–reinforced composites have some limitations, such as incompatibility with thermosets, thermoplastic, or other kinds of polymer material; poor mechanical properties; microbe infection; low temperature resistance, high moisture absorption rate, and low fire resistance. However, natural fibre–reinforced composites should meet the challenges of higher end requirements because of their high stiffness, specific functional and mechanical properties, and insulation attributes. Ecological sustainability and other legislative conditions, and the safety of the end user are all enhanced by engineering techniques. Surface treatments such as alkali, acetylation, benzoylation, enzyme, grafting, isocyanate, mercerization, methacrylate, ozone, peroxide, plasma, silane, and sodium chlorite modify the nature of the fibre or the surface of the composite as per functional requirements. Additionally, the application of nanolevel surface coatings such as fire retardancy, water repellent, oil repellent, soil repellent, anti-microbial treatment, silver halide, acrylic and other kinds of weather-resistant and heat-resistant finishes, and coloured paint on natural fibre–reinforced composites significantly helps to overcome the aforementioned challenges. Further research into selecting the right kind of fibre, processing techniques, ratio of fibre content, manufacturing methods, and technology adaptation of manufacturing machines like compression moulding and extrusion/thermoforming devices will create opportunities to solve the functional problems of Natural fibre reinforced polymer (NFRP) composites.

REFERENCES

1. Nabi Saheb, D. and Jog, J.P. Natural fiber polymer composites: A review. *Advances in Polymer Technology*, 18, 351–363, 1999.
2. Arpitha, G.R., Sanjay, M.R. and Yogesha, B. Review on comparative evaluation of fiber reinforced polymer matrix composites. *Advanced Engineering and Applied Sciences* 4, 44–47, 2014.
3. Gupta, M.K. and Srivatsa, R.K. Mechanical properties of hybrid fibers reinforced polymer composite: A review. *Polymer-Plastics Technology and Engineering* 55, 626–642, 2016.
4. Grand View Research. Automotive polymer composites market size report by resin (epoxy, polyurethane, polyamide, polypropylene, polyethylene), by application, by product, by end use, by manufacturing, and segment forecasts, 2018–2025, https://www.grandviewresearch.com/industry-analysis/automotive-polymercomposites market (accessed: 10 November 2018), 2018.
5. Ravishankar, B., Nayak, S.K. and Kader, M.A. Hybrid composites for automotive applications – A review. *Journal of Reinforced Plastics and Composites* 38(18), 835–845, 2019. doi: 10.1177/0731684419849708
6. Organization of the Petroleum Exporting Countries. *2018 OPEC World Oil Outlook 2040. Report on Medium and Long-Term Prospects to 2040 for the Global Oil Industry.* Vienna: OPEC Secretariat, 2018.
7. Sims, R., Schaeffer, R., Creutzig, F. and Xochitl Cruz-Núñez *Climate Change 2014: Mitigation of Climate Change. Report of the Intergovernmental Panel on Climate Change, Contribution of Working Group III to the Fifth Assessment.* Cambridge: Cambridge University Press, 2014.
8. Verma, D. and Sharma, S. Green biocomposites: A prospective utilization in automobile industry. In Jawaid, M., Salit, M.S. and Alothman, O.Y. (eds.). *Green Biocomposites: Design and Applications.* Cham: Springer, 2017, pp. 167–191.
9. Akampumuza, O., Wambua, P.M., Ahmed, A., Li, W. and Qin, X.H. Review of the applications of biocomposites in the automotive industry. *Polymer Composites* 38(11), 2553–2569, 2017.
10. Agarwal, J., Sahoo, S., Mohanty, S., and Nayak, S.K. Progress of novel techniques for lightweight automobile applications through innovative eco-friendly composite materials: A review. *Journal of Thermoplastic Composite Materials* February 12, 2019. doi: 10.1177/0892705718815530

11. Holbery, J. and Houston, D. Natural-fiber-reinforced polymer composites in automotive applications. *JOM* 58(11), 80–86, 2006.
12. Anandjiwala, R.D. and Blouw, S. Composites from bast fibres—Prospects and potential in the changing market environment. *Journal of Natural Fibers* 4(2), 91–901, 2007.
13. Zhao, Q. and Chen, M. Automotive plastic parts design, recycling, research, and development in China. *Journal of Thermoplastic Composite Materials* 28(1), 142–157, 2015.
14. Njuguna, J., Wambua, P., Pielichowski, K. and Kayvantash, K. Natural fibre-reinforced polymer composites and nanocomposites for automotive applications. In Kalia, S., Kaith, B.S. and Kaur, I. (eds.). *Cellulose Fibers: Bio- and Nano-Polymer Composites: Green Chemistry and Technology*. Berlin, Heidelberg: Springer-Verlag, 2011, pp. 661–700.
15. Banerjee, S. and Sankar, B.V. Mechanical properties of hybrid composites using finite element method based micro-mechanics. *Composites Part B* 58, 318–327, 2014.
16. Khan, T., Hameed Sultan, M.T.B. and Ariffin, A.H. The challenges of natural fiber in manufacturing, material selection, and technology application: A review. *Journal of Reinforced Plastics and Composites* 37(11), 770–779, 2018.
17. Saba, N., Jawaid, M., Sultan, M.T.H. and Alothman, O.Y. Green bio-composites for structural applications. In Jawaid, M., Salit, M.S. and Alothman, O.Y. (eds.). *Green Bio-Composites: Design and Applications*. Cham: Springer, 2017, pp. 1–27.
18. Black S. Looking to lighten up aircraft interiors? Try natural fibers!, https://www.compositesworld.com/blog/post/looking-to-lighten-up-aircraft-interiors—with-natural-fibers (accessed: 17 December 2018), 2015.
19. European Commission. Bio-based materials for aircraft, http://ec.europa.eu/research/infocenre/article_en.cfm?id¼4/research/headlines/news/article_18_04_26_en.html%3Finfocentre&%3Bitem¼Infocentre&%3Bartid¼48256 (accessed 15 November 2018), 2018.
20. Aly, N.M. A review on utilization of textile composites in transportation towards sustainability. *IOP Conference Series: Materials Science and Engineering* 254(4), 1–7, 2017.
21. Koniuszewska, A.G. and Kaczmar, J.W. Application of polymer based composite materials in transportation. *Progress in Rubber, Plastics and Recycling Technology* 32(1), 1–23, 2016.
22. Red, C. Composites in aircraft interiors, 2012–2022, https://www.compositesworld.com/articles/composites-in-aircraft-interiors-2012-2022 (accessed: 14 November 2018), 2012.
23. Fragassa, C. Marine applications of natural fibre-reinforced composites: A manufacturing case study. In Pellicer, E., Nikolic, D. and Sort, J. (eds.). *Advances in Applications of Industrial Biomaterials*. Cham: Springer, 2017, pp. 21–47.
24. Ansell, M.P. Natural fibre composites in a marine environment. In Hodzic, A. and Shanks, R. (eds.). *Natural Fibre Composites: Materials, Processes and Properties*. Cambridge: Woodhead, 2014, pp. 365–374.
25. Pandey, J.K., Ahn, S.H., Lee, C.S., Mohanty, A.K. and Misra, M. Recent advances in the application of natural fiber based composites. *Macromolecular Materials and Engineering* 295(11), 975–989, 2010.
26. Ashori, A. Wood-plastic composites as promising green-composites for automotive industries! *Bioresource Technology* 99(11), 4661–4667, 2008.
27. Furtado, S.C.R., Aráujo, A.L., Silva, A., Alves, C. and Ribeiro, A.M.R. Natural fibre-reinforced composite parts for automotive applications. *International Journal of Automotive Composites* 1(1), 18–38, 2014.
28. Norhidayah, M.H., Hambali, A.A., Yuhazri, Y.M., Zolkarnain, Taufik, Saifuddin, H. A review of current development in natural fiber composites in automotive applications. *Applied Mechanics and Materials* 564, 3–7, 2014.

29. Suddell, B.C. Industrial fibres: Recent and current developments. In *Proceedings of the Symposium on Natural Fibres, Rome, Italy, 20 October 2008*. Amsterdam: Common Fund for Commodities, 2008, pp. 71–82.
30. Mohammed, L., Ansari, M.N.M., Pua, G., Jawaid, M. and Islam, M.S. A review on natural fiber reinforced polymer composite and its applications. *International Journal of Polymer Science* 2015, 1–15, 2015.
31. Ramli, N., Mazlan, N., Ando Y., Leman, Z., Abdan, K., Aziz, A.A. and Sairy, N.A. Natural fiber for green technology in automotive industry: A brief review. In *The Wood and Biofiber International Conference (WOBIC 2017), Selangor, Malaysia, 21–23 November 2017*. Putrajaya, Malaysia: Institute of Physics Publishing, 2017, pp. 1–7.
32. Bharath, K.N. and Basavarajappa, S. Applications of biocomposite materials based on natural fibers from renewable resources: A review. *Science and Engineering of Composite Materials* 23(2), 123–133, 2016.
33. Zhao, D and Zhou, Z. Applications of lightweight composites in automotive industries. In Yang, Y., Xu, H. and Yu, X. (eds.). *Lightweight Materials from Biopolymers and Biofibers*. Washington: American Chemical Society, 2014, pp. 143–158.
34. Hassan, F., Zulkifli, R., Ghazali, M.J. and Azhari, C.H. Kenaf fiber composite in automotive industry: An overview. *International Journal of Advanced Engineering and Technology* 2017, 7(1), 315–321.
35. Rivera-Gómez, C., Galán-Marín, C. and Bradley, F. Analysis of the influence of the fiber type in polymer matrix/fiber bond using natural organic polymer stabilizer. *Polymers* 6, 977–994, 2014.
36. Smith, J and Bhatia, S. Natural fibers raise social issues. *Materials Today* 8, 72–81, 2005.
37. Danso, H., Martinson, D.B., Ali, M. and Williams, J.B. Mechanisms by which the inclusion of natural fibres enhance the properties of soil blocks for construction. *Journal of Composite Materials* 51(27), 3835–3845, 2017.
38. Beglarigale, A. and Yazıcı, H. Pull-out behavior of steel fiber embedded in flowable RPC and ordinary mortar. *Construction and Building Materials* 75, 255–265, 2015.
39. Danso, H., Martinson, D.B., Ali, M., and Williams, J.B. Physical, mechanical and durability properties of soil building blocks reinforced with natural fibres. *Construction and Building Materials* 101, 797–809, 2015.
40. Papaspyrides, C.D., Pavlidou, S., and Vouyiouka, S.N. Development of advanced textile materials: Natural fibre composites, anti-microbial, and flame-retardant fabrics. *Proceedings of the Institution of Mechanical Engineers Part L: J. Materials: Design and Applications* 223, 2009, 91–102. doi: 10.1243/14644207JMDA200
41. Dweib, M., Hu, B., O'Donnell, A., Shenton. H. and Wool, R. All natural composite sandwich beams for structural applications. *Composite Structures* 63, 147–157, 2004.
42. Riedel, U. and Nickel, J. Applications of natural fiber composites for constructive parts in aerospace, automobiles and other areas. In A. Steinbüchel (ed.). *Biopolymers (General Aspects and Special Applications)*. New Jersey: John Wiley & Sons, Inc., 2003, Vol. 10, pp. 1–11.
43. Zakriya, G.M. & Ramakrishnan, G. 2019. Jute and hollow conjugated polyester composites for outdoor & indoor insulation applications. *Journal of Natural Fibres* 16(2), 185–198. doi: 10.1080/15440478.2017.1410515
44. Hu, N. *Composites and Their Applications*, IntechOpen, 2012. doi: 10.5772/3353

10 Applications of Composites in Artefact and Furniture Making

10.1 INTRODUCTION

Researchers have welcomed the move to impose guidelines for ensuring a healthier and safer environment, and have begun research into creating new ideas in eco-composite technology[1-3] An eco-composite can be defined as a composite with better eco-friendly features and ecological advantages over synthetic or conventional composites. Eco-composites can be manufactured from natural fibres or by blending various natural fibres with polymers and polymer derivative matrices. The increasing cost of raw materials used to produce engineering and standard plastics, the future sustainability of natural resources, and threats to the environment have made it essential to use natural raw materials to develop and fabricate polymer composites.[4,5] In recent years, the reinforcement industry had been dominated by the use of synthetic fibres. The versatility of natural fibre reinforcement gained much attention and interest in substituting them for synthetic fibres in various applications. Researchers and manufacturers exploited both softwoods and hardwoods to extract the fibres for reinforcement in various composite manufacturing processes.[6]

To avoid the depletion of forests and natural ecosystems, agricultural crops have been utilized to carry out research and development on the commercialization of natural fibre polymer composites. Bamboo, jute, kenaf, sisal, hemp, cotton, and coir are some of the agricultural crops that can be reinforced with polymers.[7] In some developing countries, crop fibres play a vital role in economic growth, for example, sisal in Tanzania, jute in Bangladesh, and cotton in some West African countries. Blending of natural fibres with polymer matrices from both renewable and nonrenewable sources is used to fabricate polymer composites that are competitive with synthetic composites, and these have been gaining attention over the last decade.[8] Bio-based polymers and biodegradable plastic products from renewable sources are used to produce sustainable and eco-friendly composite products, which can compete in and capture the current market, which is dominated by petroleum-derived polymers.[9]

Sustainable and biodegradable biocomposites are produced from bamboo fibres as a substitute for glass fibre composites for outdoor and indoor applications. Bamboo fibre–reinforced composites have a high strength to weight ratio, durability, dimensional stability and pliability into complex shapes. The reinforcement of bamboo fibres using a hybridization technique with thermoplastic, thermoset, jute, oil palm, coir, and other fibre blends provides high mechanical and thermal insulation properties.[10] Sustainable bamboo and other agro-crop plants evolved to become the backbone for the socioeconomic status of society.[11,12] Adhesive wear, frictional performance, and higher sliding velocity are good in bamboo and other crop fibre–reinforced epoxy composites.[13] Pineapple-leaf fibre, bamboo fibre, and banana fibre reinforced with polypropylene matrix composite show the highest flexural properties.[14] Its strength and rigidity, the inferior microfibrillar angle to its fibre axis, and its thicker cell wall are reasons to consider bamboo fibre as 'nature's glass fibre'.[15]

The nonabrasive quality of jute fibre–reinforced composites in the mould form results in higher fibre loading and less pace of damage compared to glass fibre.[16] It degrades at more than 200°C, and within this limitation, thermoset, thermo plastic, and natural binding material may be added to produce jute molds.[17] Therefore, in terms of a substitute for wood, jute fibre–reinforced composites would be an ideal solution. Jute-reinforced composites represent good value for customers without compromising anything in terms of their properties.[18] After consequent loading cycles, permanent distortion of natural fibre preforms is estimated to be higher than with synthetic fibre preforms.[19] Jute fibre layers bonded with phenolic resin have a wood-like finish. Poor compatibility between the polymer matrixes with hydrophilic fibres leads to weak interfaces and affects the mechanical properties of composites.[20] Surface modification techniques are carried out on the composite material to improve the interfacial adhesion by adding coupling agents and subjecting it to a pelletizing process.[21]

To avoid the use of resin in making composites, some other mechanical bonding is required to impart knots or bond to interlock the hybrid fibre. Nonwoven fabric with randomized fibre arrangement possesses good mechanical strength,[22] and needle punching technology is well suited for jute fibre. Jute fibres more than 80 mm long had good fibre lock at mechanical needling. The water absorption behaviour of fibre present in the composite structure leads to swelling and dimensional instability. If this is to be overcome, adding coatings or finishes to composites material is essential.[22] Palm kernel fibre–reinforced boards can be used as an alternative to timber for construction work and for making furniture.[23] Researchers have shown that jute-reinforced polymer composite with red mud/fly ash composite is the most promising substitute for timber by proving its physical, chemical, mechanical, weathering, and fire resistance properties.[24]

10.2 ARTEFACTS MOULD MAKING

Compare to conventional composites and traditional wood, reinforced bamboo composites have been developed that show good dimensional stability, durability, and weather resistance and high impact resistance, are low maintenance, nontoxic and flame resistant. A blend of 70% recycled bamboo fibres/origin fibres and 30% recycled high-density polyethylene (HDPE) gives a composite with special characteristics like termite resistance, high thermal stability, and shape retention, suitable for making artefacts like pottery vessels, buttons, utensils, deck tiles, railings, dustbins, decking accessories, laminated bamboo lumber for structural applications, and so on.[25,26] A blend of 44% bamboo fibre reinforced with epoxy resin gives a composite that shows excellent wear resistance.[27]

Artefacts with moderate mechanical properties can be produced using reinforced fibre preforms and nonwoven mats in a compression moulding process or loose fibres or blended fibres in an injection moulding process. Dies were made in the shape of the final artefacts.[27] The hand lay-up process is the most common method for manufacturing large and complex products. This method requires minimum equipment and inexpensive moulds made of wood, reinforced plastic, plaster of Paris, and so on. One mould, either male or female, is sufficient to produce the articles. The end products have finish on one side, which is the side in contact with the surface of the die mould. Commonly, polyester and epoxy resins are used. The moulding process is as follows: i) application of release agent to the surface of the mould, ii) application of gel coating to the surface of the fibre, iii) carrying out the lay-up operation as per requirement determined by the g/m² or thickness of the product, iv) curing and releasing the mould to take the product out of the die, v) trimming the product to enhance the appearance, and vi) painting on functional finishes as per the end use of the product.[28,29]

Rotational moulding, also called *roto moulding/roto casting*, is depicted in Figure 10.1. It is used to produce hollow reinforced-fibre plastic products. Rotational moulding has benefits in terms of its moderately low level of residual stress and the fact that moulds are inexpensive. It is possible to produce a hollow composite product more than 2 m³ in size in one piece of wrap. This process is used to produce

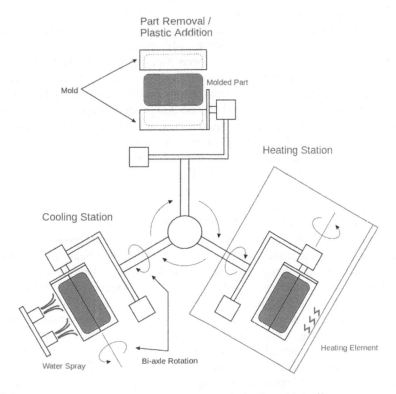

FIGURE 10.1   Rotational moulding machine to make hollow objects.[30]

toys, leisure crafts, highly aesthetic fibre-reinforced products, water tanks, and complex medical products. Various kinds of resins can be used in this machine, such as acetate butyrate, polyamide, polyurethane, polypropylene, ethylene vinyl acetate and plastisols, polyethylene, fluorocarbons, polycarbonate, elastomers, nylon and cross-linked polyethylene.[30]

The process of rotational moulding of fibre reinforced plastics is begun by feeding a known amount of fibre/wood dust/plastic in granular, powder, or viscous liquid form into a hollow shell mould. The mould is rotated or rocked about the two main axes at reasonably low speeds, and the poured and enclosed fibre solution is heated so that it adheres to the mould shape and forms a uniform layer against the mould surface. The mould rotates continuously up to the cooling phase, so that the material adhering in the mould retains its required final shape as it solidifies. When the material is sufficiently rigid, a water spraying or ventilation cooling process takes place and mould rotation is stopped. The formed fibre-reinforced plastic product or composite is removed from the mould and continues the process for bulk production.[30]

10.3   FURNITURE MAKING

Natural fibre–reinforced composite material can replace wood and timber in making furniture, but this involves the innovative redesign of the components. Innovation is an essential pillar in redesigning the product to meet a wide choice of user requirements related to cost, shape, visual appearance, and quality.[31] To remain competitive in this market, manufacturers need to develop innovative solutions to meet the needs of customer or consumer.[32,33] Since 1948, thermoset and thermoplastic synthetic polymers with natural fibres have been used in the furniture industry. Hemp and flax fibre–reinforced composites have also been exhibited at furniture fairs.[34,35] However, green composites have not been commercially successful and awareness of this product is nonexistent in the market. The design specifications of a piece of furniture should conform to legislative requirements and it should meet accepted testing standards such as ISO/ASTM/BIS. Economic requirements such as manufacturing cost, assembly cost, and finally storage and transportation cost need to be considered, and a safety standard established for the enhanced reliability of the product.[36]

An increase in cellulose content in the fibres clearly correlates with an increase in stiffness and strength. In addition, cellulose crystallinity, the crystallite aspect ratio, and the microfibril direction affect the stiffness of the cell wall.[36] Wood fibre preforms from pulp fibres and resin-impregnated fibre mats can be put through compression moulding or resin transfer moulding processes to make components of furniture.[37] In the compression moulding technique, the natural fibres and the thermoplastic fibres should have similar dimensions in order to maintain efficient mixing uniform hydrodynamic properties. This mixture is compressed in a hot press to form a mat with a flat pattern. Complex patterns can also be made in a double-curvature hot pressing machine. After making a fibrous composite as per the requirements of the design, a trimming process is carried out using appropriate tools.[38] The estimated stiffness offered by this kind of cellulose fibre composite will be in the range of 20–90 GPa.[39] Banana fibre–reinforced epoxy resin composite is considered a suitable material for making furniture. A schematic view of the furniture-making process is shown in

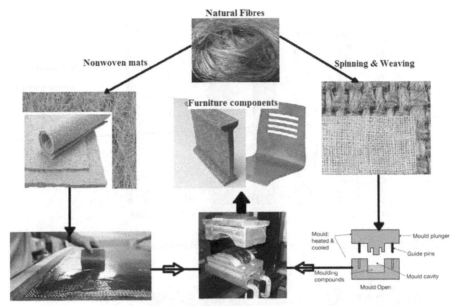

FIGURE 10.2   Schematic view of furniture-making process.

Figure 10.2. By means of computational analysis, banana fibre–epoxy composite mechanical properties are shown to be good and can be significantly improved by increasing the amount of banana fibre in the polymer matrix.[40]

Bioplastics offer a range of functional composites to the furniture industry. Advanced biopolymers such as PHAs (polyhydroxyalkanoates) and PLA (polylactic acid) matrix with a variety of natural fibres provides biodegradable composites or bioplastics. PLA is versatile and has excellent barrier qualities, and it is available in different grades for commercial usage. The PLA matrix replaces the need for ABS (acrylonitrile butadiene styrene), PP (polypropylene), and PS (polystyrene). [41] To process PLA in the processing equipment, slight and simple adjustments need to be done as per functional requirements. Among the 360 million tonnes of plastic production, bioplastics or biocomposites represent only around 1%. But as per rising demand for eco-friendly products, the market is growing steadily. Around 3,00,000 highly skilled technicians will be required in this composite industry by the year 2030.[41] The future market for bioplastic or biocomposites and their market segment are depicted in Figures 10.3 and 10.4 respectively.

10.4   MACHINERY DEVELOPMENT AND SENSORS UTILIZATION

Near-net-shape pin tooling (NPT) and subtractive pin tooling (SPT) are two separate reconfigurable systems that help to align the machine beds by adjusting square pins. Pins and sensors can be used to accommodate dimensional changes in a composite part; they feature integrated heating/cooling channels that allow the tool to be

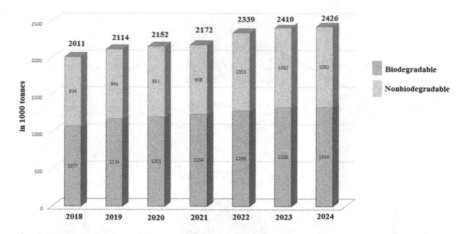

FIGURE 10.3   Forecast of future requirements for bioplastics.[41]

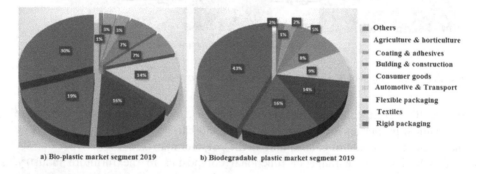

a) Bio-plastic market segment 2019   b) Biodegradable plastic market segment 2019

FIGURE 10.4   Market segments of bioplastic.[41]

heated from an ambient temperature to 800°F (427°C) and then back to ambient in 12 minutes. Using computerized temperature controls, technicians can cure parts to a specific schedule and can vary the amount of heat across the tool. Tools used for light resin transfer moulding (light RTM) equipped with specialized 5-axis milling and pattern building features. Low-cost rapid tooling uses patented metal deposition technology to help form the shape of the product with faceplate tools. Electron beam metal deposition processes generate near-net-shape 3D geometries that meet aerospace quality standards for titanium, nickel, and steel alloys.[42]

Pinette P.E.I. composite press systems improve composite equipment efficiency, accuracy, and reliability through the following advantages:[43]

- Press force from 10 to 100,000 kN
- Heating platen size up to 8 metres
- Active levelling control
- Heating process up to 500°C
- Homogenous temperature with accuracies up to ±3°C on the heating platen surface

- Heating (oil, electric) and cooling platens
- Easy access for automation tasks or maintenance
- Customized equipment solutions
- High level of automation

A fully automated resin transfer moulding (RTM) machine is used for aerospace and automotive manufacturing applications. This machine is designed to increase moulded part throughput and improve process consistency and is able to produce up to 100,000 parts a year. It has the following features:[43]

- Injection machine
- Tool integration
- Hydraulic press up to 10,000 kN
- Platen size up to 3,000 × 2,000 mm
- Heating and cooling platens
- Integrated heating power unit
- Upper moveable table for mould loading with covered guard insulation along with trap for injection
- Automatic shuttle transfer system
- Temperature uniformity up to 300°C
- 6-axis robot
- In touch supervision software with full traceability

10.5 CONCLUSION

Bamboo, banana, jute, kenaf, hemp, sisal, oil palm and recycled fibres are extensively used in composite industries for socio-economic development and empowerment of peoples. Cost-effective and eco-friendly biocomposites are popular for their functionality. Composites are changing the world and creating many wonderful opportunities in fields like furniture, flexible and rigid packaging, building and construction, and the automobile sector. Innovative demands created by this sector[44] need to be addressed, and the choice of effective design and versatility in the changing matrix and moulding process is essential.[45] Appropriate selection of fibres, appropriate utilization of textile forming (preform) technology, and emerging adaptation of controllers and sensors in machinery will help to create fault-free, quality fibre-reinforced composites.

REFERENCES

1. Gaceva, G.B., Avella, M., Malinconico, M., Buzarovska, A., Grozdanov, A., Gentile, G. and Errico, M.E. Natural fibre eco-composites. *Polymer Composites* 28, 98–107, 2007.
2. Selke, S. *Biodegradation and Packaging* (2nd ed.). Surrey, UK: Pira International Reviews, 1996.
3. Avella, M., Bonadies, E., Martuscelli, E. and Rimedio, R. European current standardization for plastic packaging recoverable through composting and biodegradation. *Polymer Testing* 20, 517–521, 2001.
4. Jawaid, M. and Abdul Khalil, H.P.S. Cellulosic/synthetic fibre reinforced polymer hybrid composites: A review. *Carbohydrate Polymers* 86, 1–18, 2011.

5. Puglia, D., Biagiotti, J., Kenny, J.M. A review on natural fibre-based composites – Part II. *Journal of Natural Fibers* 1, 23–65, 2005.
6. Chand, N. and Fahim M. *Tribology of Natural Fiber Polymer Composites.* Cambridge, UK: Woodhead Publishing Ltd., 2008.
7. Coutts, R.S.P., Ni, Y. and Tobias, B.C. Air-cured bamboo pulp reinforced cement. *Journal of Materials Science Letters* 13, 283–285, 1994.
8. Jawaid, M., Abdul Khalil, H.P.S. and Abu Bakar, A. Mechanical performance of oil palm empty fruit bunches/jute fibres reinforced epoxy hybrid composites. *Materials Science and Engineering A* 527, 7944–7949, 2010.
9. Mohanty, A.K., Misra, M. and Drzal, L.T. Sustainable bio-composites from renewable resources: Opportunities and challenges in the green materials world. *Journal of Polymers and the Environment* 10, 19–26, 2002.
10. Jawaid, M. and Abdul Khalil, H.P.S. Cellulosic/synthetic fibre reinforced polymer hybrid composites: A review. *Carbohydrate Polymers* 86, 1–18, 2011.
11. Shin, F.G. and Yipp, M.W. Analysis of the mechanical properties and microstructure of bamboo–epoxy composites. *Journal of Materials Science* 24, 3483–3890, 1989.
12. Okubo, K., Fujii, T. and Yamamoto, Y. Development of bamboo-based polymer composites and their mechanical properties. *Composites Part A: Applied Science and Manufacturing* 35, 377–383, 2004.
13. Nirmal, U., Hashim, J. and Low, K.O. Adhesive wear and frictional performance of bamboo of fibres reinforced epoxy composite. *Tribology International* 47, 122–133, 2012.
14. Chattopadhyay, S.K., Singh, S., Pramanik, N., Niyogi, U.K., Khandal, R.K., Uppaluri, R. and Ghoshal AK. Biodegradability studies on natural fibres reinforced polypropylene composites. *Journal of Applied Polymer Science* 121, 2226–2232, 2011.
15. Li, S.H., Zeng, Q.Y., Xiao, Y.L., Fu, S.Y. and Zhou, B.L. Biomimicry of bamboo bast fibre with engineering composite materials. *Materials Science and Engineering: C* 3, 125–130, 1995.
16. Chandan, D., Basu, D., Abhishek, R. and Amarnath, B. Mechanical and dynamic studies of epoxy/VAc/HMMM IPN-jute composite systems. *Journal of Applied Polymer and Science* 91, 958–963, 2003.
17. Karaduman, Y., Gokcan, D., and Onal, L. Effect of enzymatic pretreatment on the mechanical properties of jute fiber-reinforced polyester composites. *Journal of Composite Materials* 47, 1293, 2013.
18. Gon, D., Das, K., Paul, P. and Maity, S. Jute composites as wood substitute. *International Journal of Textile Science* 1(6), 84–93, 2012. doi: 10.5923/j.textile.20120106.05
19. Francucci, G., Rodrı́guez, E.S. and Vázquez, A. Experimental study of the compaction response of jute fabrics in liquid composite molding processes. *Journal of Composite Materials* 46, 155, 2012.
20. Gon, D., Das, K., Paul, P. and Maity, S. Jute composites as wood substitute. *International Journal of Textile Science* 1(6), 84–93, 2012.
21. Akdogan, A. and Vanli, A.S. Material characterization of different dimensioned wood particle-reinforced polymer composites. *Journal of Thermoplastic Composite Materials* 26(9), 1237–1248, 2013.
22. Gibson, R.F., Chaturvedi, S.K. and Sun, C.T. Complex moduli of aligned discontinuous fiber reinforced polymer composite. *Journal of Material Science* 17, 3499–3509, 1982.
23. Adedeji, Y.M.D. and Ajayi, B. Cost-effective composite building panels for walls and ceilings in Nigeria. In *11th International Inorganic Bonded Fibre Composite, Conference on Nov 05–07,* Madrid – Spain, 2008, pp. 151–159.
24. Saxena, M., Morchhale, R.K., Asokan, P., and Prasad, B.K. Plant fiber – Industrial waste reinforced polymer composites as a potential wood substitute material. *Journal of Composite Materials* 42(4), 367–384, 2008.

25. Mahdavi, M., Clouston, P.L. and Arwade, S.R. A low-technology approach toward fabrication of Laminated Bamboo Lumber. *Construction and Building Materials* 29, 257–262, 2012.
26. Abdul Khalil, H.P.S., Bhat, I.U.H., Jawaid, M., Zaidon, A., Hermawan, D. and Hadi, Y.S. Bamboo fibre reinforced biocomposites: A review. *Materials and Design* 42, 353–368, 2012.
27. Nirmal, U., Hashim, J. and Low, K.O. Adhesive wear and frictional performance of bamboo of fibres reinforced epoxy composite. *Tribology International* 47, 122–133, 2012.
28. Madsen, B. and Gamstedt, E.K. Wood versus plant fibers: Similarities and differences in composite applications. *Advances in Materials Science and Engineering* 2013, 564346, 2013. doi: 10.1155/2013/564346
29. http://www.dcmsme.gov.in/publications/pmryprof/chem/ch12.pdf (accessed on 28/03/2020)
30. http://www.dhrotomold.com/news/what-is-rotational-molding
31. Diego-Mas, J.A. and Alcaide-Marzal, J. Single users' affective responses models for product form design. *International Journal of Industrial Ergonomics* 53, 102–114, 2016.
32. Ulrich, K.T. and Eppinger, S.D. *Product Design and Development*, 2. New York: McGraw-Hill, 2000.
33. Cavallucci, D. and Lutz, P. Intuitive design method. A new approach on design methods integration. In *Proceedings of ICAD First International Conference on Axiomatic Design*, 2000, pp. 21–23.
34. Pil, L., Bensadoun, F., Pariset, J. and Verpoest, I. Why are designers fascinated by flax and hemp fibre composites? *Composites: Part A* 83, 193–205, 2016.
35. http://www.Designmuseumgent.Be/ENG/Exhibitions-2015/Synthetic-By Nature Php (2015)
36. Thygesen, A., Thomsen, A.B., Daniel, G. and Lilholt, H. Comparison of composites made from fungal defibrated hemp with composites of traditional hemp yarn. *Industrial Crops and Products* 25(2), 147–159, 2007.
37. Almgren, K.M., Gamstedt, E.K., Nygard, P., Malmberg, F., Lindblad, J. and Lindstrom, M. Role of fibre-fibre and fibre-matrix adhesion in stress transfer in composites made from resin impregnated paper sheets. *International Journal of Adhesion and Adhesives* 29(5), 551–557, 2009.
38. Lindström, M., Berthold, F., Gamstedt, K., Varna, J. and Wickholm, K. Hierarchical design as a tool in materials development. In *Proceedings of the 10th International Conference on Progress in Biofiber Plastic Composites*, Toronto, Canada, 2008, pp. 1–7.
39. Mehmood, S. and Madsen, B. Properties and performance of flax yarn/thermoplastic polyester composites. *Journal of Reinforced Plastics and Composites* 31, 62–73, 2012.
40. Ishak, M.I, Ismail, C.N., Khor, C.Y., Rosli, M.U., Jamalludin, M.R., Hazwan, M.H.M., Nawi, M.A.M., and Mohamad Syafiq A.K. Investigation on the Mechanical Properties of Banana Trunk Fibre–Reinforced Polymer Composites for Furniture Making Application. In *IOP Conf. Series: Materials Science and Engineering*. IOP Publishing. 2019, Vol. 551, p. 012107
41. European bioplastics, nova – Institute (2019), www.european-bioplastics.org/market/ (accessed: 29/03/2020)
42. https://www.materialstoday.com/composite-applications/features/mold-and-tooling-advances-promising-for/ (accessed: 29/03/2020)
43. https://pinetteemidecau.eu/en/hydraulic-presses/rtm-hp-rtm (accessed: 29/03/20)
44. http://www.3mb.asia/composite-materials-in-the-production-of-furniture/
45. Filip, I. Ciupan, E Cionca, I. Ciupan, M. Pop, E. Campean, E. Heres, V. Rat, F Gherghel, C. *Proceedings of the XXIth Int. Scientific Conf. "INVENTICA 2017,* 38–45, (2017)

11 Recycling of Natural Fibre Composites

11.1 INTRODUCTION

The day-to-day process of environmental regulation involves many preventive measures. The environmental impact of the disposal of natural fibre–reinforced composite materials in landfills is an issue because of the industrial scale of commercialization. Landfill is a quite inexpensive disposal method but it is the least favoured waste management option.[1] Mechanical recycling, chemical recycling, and thermal recycling are used to separate the fibres and resins from the used composite material or the scrap produced during composite manufacturing. Industrial requests for products using recycled fibres and resins are still rare or partial because reclaimed or recycled fibre composites are considered to be of lower quality compare than virgin fibre-reinforced composites.[2] Recycled fibres are not fully controlled in terms of length, fineness, length distribution, poor surface adhesion, and grade variations on batch production.[2] The unsatisfactory quality factors can be eliminated by adopting appropriate technology for processing recycled fibres.

Recycling operations involve some conditions such as temperature, pressure, equipment volume, and toxicity of catalyst, and solvents may affect or determine the recycled fibre's end use. A comprehensive assessment must be carried out in order to categorize these diverse technologies in terms of their efficiency, environmental impact, and commercial viability.[3] Three types of recycling process are available, categorized as primary, secondary, and tertiary. In a primary recycling process, the recyclable fibre or resin is recovered from the composite and reused for the purpose of producing the parent or same composite product. In the secondary recycling process, the recyclable fibre or resin is reused in some other way, without reprocessing, to produce other products. In the tertiary recycling process, the composite material or product is chemically altered in order to make it reusable.[4]

Furthermore, in a universal and eco-friendly approach to the recycling process, one must examine and consider the end of life of reinforced composite products at an early stage in their development process.[5] The recycling process for reinforced composite materials and products should include end-of-life strategies and logistics at the product development stage.[6] Design and process engineers must balance energy efficiency, safety, cost, and need when thinking about how the composite product will be dealt with at the end of its useful life.[7] The common characteristics of design reuse strategies are:[8–10]

- **Recoverability:** This measures whether the product can be recovered or not after use.
- **Functionality:** This is the average outstanding useful lifespan of a product at end-of-use compared to its premeditated lifetime.

- **Technological maturity:** This is the degree of technical amendment adopted or the phases of the inclusion of new features in the product.
- **Components:** This is the number of unique parts that can be made and adopted to various kinds of products.
- **Level of integration:** This is the independence of product components or level of contribution needed to assure the functionality of a particular end product.

11.2   CHEMICAL RECYCLING

Chemical recycling, also known as *feedstock* or *tertiary* recycling, uses a high-temperature process to break down the structural bonds in the natural fibre–reinforced polymer composite to be recycled.[11] Chemical recycling encompasses diverse procedures such as pyrolysis, hydrocracking, hydrolysis, and gasification.[12] Pyrolysis is carried out in an oxygen-lean atmosphere to break down the large molecules into smaller ones through the influence of heat. Pyrolysis is appropriate for use in dealing with mixed waste or residuals of automotive parts produced after shredding. The pyrolysis process converts plastics into gases, a varied liquid of hydrocarbons, and solid char.[12]

Some high molecular weight fibre-reinforced polymers, due to the absence of hydrogen and oxygen, are refined into high-value-added petrochemical feedstock as listed in Table 11.1.

Hydrocracking is the alteration of higher boiling point chemicals in crude oil to low boiling point, such as gasoline or jet fuel, using hydrogen under high pressure. Gasification involves the introduction of carbon, comprising material from coal and biomass, using a controlled amount of air or oxygen. The molecules are thereby broken down into hydrogen, carbon monoxide, and other gases.[14] Hydrolysis process also helps to break down the large molecule compounds into smaller ones in the presence of water. An example of the hydrolysis process is the depolymerization of condensation polymers such as PU, PLA, and PET. The monomers that result can be reused in the synthesis of new polymers.[15]

Solvolysis is a process that involves chemical treatment by means of a solvent to degrade the resin or matrix of the reinforced composite. This technique is applied

TABLE 11.1
Recycling natural fibre–reinforced composites out of pyrolysis[12,13]

Polymer	High-temperature products	Low-temperature products
PA-6	Aramid (aromatic polyamides)	Caprolactam
PE	Gases, light oils	waxes, paraffin, oils, α-olefins
PET	TFE	Benzoic acid, vinyl terephthalate
PTFE	TFE	Monomers
PMMA	Tetrafluoroethylene (TFE)	Methyl methacrylate (MMA)
PP	Gases, light oils	Vaseline, olefins
PS	Styrene	Styrene
PVC	Toluene	HCl, benzene

to unsaturated polyesters in sheet moulding composites. Hydrolysis takes place between 220°C and 275°C with or without added solvent/catalyst degrade into glycols, carboxylic acids and a styrene–fumaric acid as copolymer. Different conditions and solvents are used to recycle thermoplastic and thermoset fibre-reinforced composites.[16] Alkaline catalysts include potassium hydroxide (KOH) or sodium hydroxide (NaOH). Acidic catalysts are purposely used to degrade high resistant resins like epoxy resins at low temperatures. Methanol, propanol, ethanol, and glycols are some other solvents that can be used with or without additives/catalysts.[17]

Low temperature and pressure (LTP) solvolysis is commonly carried out below 200°C at atmospheric pressure. Catalysts and additives are generally required to degrade the resin present in reinforced composites at low temperature along with a stirring action. Sulphuric acid, acetic acids, and nitric acid are used as the medium. In high temperature and pressure (HTP) solvolysis (at more than 200°C), alkaline conditions are used to recover the epoxy monomers. The solutions used in the LTP process cannot be used again and are difficult to recycle.[18] Common recycling processes for fibre-reinforced polymer methods are depicted in Figure 11.1.

11.3 THERMAL RECYCLING OF THERMOSET AND THERMOPLASTIC COMPOSITE

Thermal processes include fluidized-bed pyrolysis, pyrolysis, and pyrolysis assisted with microwaves called thermal fluid. These methods involve the recovery of fibre. Ultimately it also recovers the inserts, fillers, gases such as hydrogen, carbon dioxide, methane, oil fraction and char of fibres. Residual resin is volatilized into lower-weight molecules. The thermal processes work between 450°C and 700°C depending on the resin matrix utilized for reinforcement. Lower temperatures are also adapted

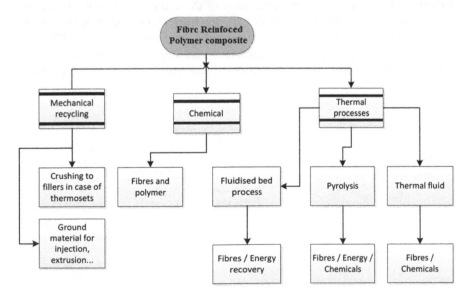

FIGURE 11.1 Recycling methods for fibre-reinforced polymer.[15]

for polyester resins, epoxides, and thermoplastic resins.[19–21] The fibres are tainted by produced char and they require a post-treatment process at a minimum furnace temperature level of 450°C to burn it off.[22] Recycled fibre's tensile strength is reduced by over 30% compared to virgin fibre. The properties of recycled fibre depend on their different sensitivity to pyrolysis conditions.[23]

11.4   MECHANICAL RECYCLING

Mechanical recycling consists of milling and grinding the fibre-reinforced composite materials to a fine texture. After an initial crushing or shredding stage, it is further ground into smaller tuft pieces. Recycling machines sieve different sized pieces out of the material sieving into resin-rich powders and fibres. Various cut lengths, close roller settings determine the recycle fibre quality. Recycled fibre can be used as filler or reinforcement. Mechanical recycling of thermoplastic fibre composite effects in outputs, undergo the consequent hot moulding/shear processing like injection moulding/extrusion. Mechanical recycling entirely categorized according to the anticipated level of fibre reinforced composite adulteration and to the composite product life cycle.[24] Mechanism of mechanical recycling shown in Figure 11.2.

11.5   THERMOSET AND THERMOPLASTIC SEPARATION

A matrix removal process has been developed for thermoset matrix composite materials. It consists of dissolving the resin phase in water under supercritical conditions. At a pressure in the range of 200 bars and at a temperature of 400°C, water acts like a real solvent to remove the resin phase. After this, the dry fibre reinforcement remains as per the original composite structure with slight fibre degradation. In thermoplastic composites, the resin part of the composite cannot be reformed again by heating. Thermoset composites are deposited in a special machine for processing. Processed reinforcing fibres are separated out from the resin and filler part. The recycled fibres can be reused as reinforcing medium in other applications. The recycled resin and filler part is continuously used again as filler for numerous end

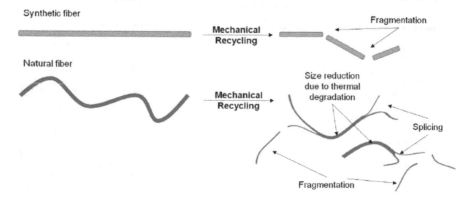

FIGURE 11.2   Mechanical recycling mechanism.[15]

uses.[25] The oil recuperated from the degradation of polyester resin contains monomers that can be reused as a new resin to provide a cost-effective solution.[26] Recycled thermoplastic composites are put through a grinding process which converts them into small particles; these can be fed into an injection moulding machine together with virgin thermoplastic materials to make new products.

11.6 BIOLOGICAL RECYCLING

Biological recycling is a process of degrading large molecules into smaller ones. Bacteria, fungi, and microorganisms assist in the polymer composite degradation process with the help of water and air. Biosynthesized composite polymers such as chitin, cellulose, and PLA (polylactic acid) are biodegraded under the combined effect of pressure, microorganisms, and P_H value and composted under different processing conditions. This kind of biodegraded recycled fibre is returned to the production cycle.[27–29] Plastics like PBS (polybutylene succinate) and PCL (polycaprolactone) are made out of petrochemical products but they are biodegradable in nature. Acetyl cellulose, PE, and nylon are considered as nonbiodegradable although they are all made from biomasses. However, PE, which has widespread uses, can initially be degraded using abiotic factors such as temperature and UV light by the oxidation process of polyethylene chains. After that, the size of polyethylene molecules drops to an adequate range for enzymatic exploitation, normally from 10 to 50 microns. Similarly, acetyl cellulose composite can be altered into degradable material based on acetylation grade[30] (Figure 11.3).

11.7 PURIFICATION

The purity of the recycled fibre surface has an impact on its ability to adhere with new resin. Previous resin residues remain on the surface after the pyrolysis or solvolysis process. The single fibre strand tensile strength seems to improve, even though, at the process of reinforcement or reincorporation into a new composite, old residue reduces adherence ability with the new matrix, leading to poorer mechanical properties. [31,32]

Primary mechanical recycling is partially suited to uncontaminated thermoplastic reinforced composites. Recycling of reinforced composites material does not require preparative procedures such as separation or purification. Thermal stability

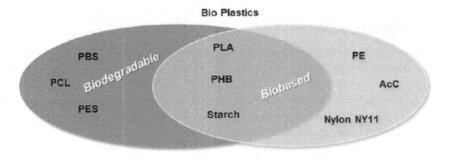

FIGURE 11.3 Biodegradable and nonbiodegradable plastic composite.[15]

is crucial for recycled fibre reinforced composites for primary mechanical recycling. Hence, polyolefins like PP, PE, PVC, and PET are the most popular mechanical recycling method for thermoplastic composites.[33,34]

Secondary mechanical recycling is carried out for used products such as plastic bags, packages, and detergent containers. Generally, secondary recycling requires a further purification process. After the mechanical process, solvents like petroleum ether and toluene are employed to dissolve the polymer matrix and re-precipitate it one more time. The prerequisite procedure of purification leads to chain scissions and lower molecular weight, reducing the mechanical properties. The efficiency of the refining process deals with the level of impurity in the main polymer by another one as sample of PET in PVC. There is a change in the fragrance and colour of the recycled fibre material. Certain additives are suggested to uphold the chain length and achieve superior properties in the finished recycled product. To enhance the material's long-term stability, light stabilizers such as Recyclossorb 550 are added to PE pellets and HDPE applications. Antioxidants are used for polyolefin recycling to retain constant melt-flow rate and thus continue a stable flow behaviour.[35-37]

11.8   RE-MOULD OR DEVELOP A PRODUCT

Various kinds of categorizations must be carried out before and after the recycling process. Afterwards, the recycled fibrous material or re-processed semi-product has to be confirmed through quality testing and need to enrich composite design or end product design by adequate space for the designer. Specific tests must be developed to meet relevant and reliable standards. A multi-level approach should also be considered to develop new recycling processes and methods towards its industrialization and feasible adaptation. Based on the test results of recycled fibre or resin data, the sequence of re-mould production assembly must be identified. For example, needle-punching technology is an appropriate choice for processing all kinds of fibres and resin fibre together, the needle punched nonwoven is further processed relevant to the requirement of product design; moulding process can be done on the nonwoven web by hot compression moulding technique. Research and development issues in terms of compression moulding machinery, such as uniform heat distribution, ensuring the temperature level at the inside of processing product during the manufacturing process, using heat flux sensors, and the pattern of design flexibility on the mould, are all essential to remould or develop products using recycled fibres.

11.9   CONCLUSION

Recycling options and limitations can be included from early on in the product design stages to the end-of-life of the product. Skills and competences relevant to the recycling process are analyzed via eco-design characteristic strategies. Developing a strategy to improve the low mechanical attributes of recycled fibres by proper analysis is recommended for product diversification. Generally, recycling processes always have the following environmental benefits:

- Decreasing landfill space requirements and encouraging waste management
- Lessening the usage of virgin fibres and conserving resources

- Reducing the consumption of energy and water
- Avoiding illness caused by pollution
- Reducing the cost of manufacturing

REFERENCES

1. Jacob, A. Composites can be recycled. *Reinforced Plastics* 55, 45–46, 2011.
2. Pimenta, S. and Pinho, S.T. Recycling carbon fibre reinforced polymers for structural applications: Technology review and market outlook. *Waste Management* 31, 378–392, 2011.
3. Oliveux, G., Dandy, L.O. and Leeke, G A. Current status of recycling of fibre reinforced polymers: Review of technologies, reuse and resulting properties. *Progress in Materials Science* 72, 61–99, 2015.
4. Recyling Consortium. Primary, Secondary and Tertiary Recycling Explained. https://www.recyclingconsortium.org.uk/primary/, 2014 (accessed: 30/03/2020).
5. Perry, N., Pompidou, S., Mantaux, O. and Gillet, A. Composite fiber recovery: Integration into a design for recycling approach. *Technology and Manufacturing Process Selection*, Springer, pp. 281–296, 2014. ff10.1007/978-1-4471-5544-7_14ff. ffhal-01066975f
6. Gaustad, G., Olivetti, E. and Kirchain, R. Design for recycling: Evaluation and efficient alloy modification. *Journal of Industrial Ecology* 14(2), 286–308, 2010. doi: 10.1111/j.1530-9290.2010.00229.x
7. Vallet, F., Millet, D. and Eynard, B. How ecodesign tools are really used. In *Requirements List for a Context-Related Ecodesign Tool. CIRP Design Conference Proceeding*, Nantes, 2010.
8. Nick, M. (2006). Presentation of "The Potential of Remanufacturing to Increase Resource Efficiency", Japan, 2006.
9. Rose. Design for Environment: A method for formulating product end-of-life strategies. http://www.seas.columbia.edu/earth/RRC/documents/2000.dfe.diss.rose.pdf, 2000
10. Brems, A., Baeyens, J. and Dewil, R. Recycling and recovery of post-consumer plastic solid waste in a European context. *Thermal Science* 16(3), 669–685, 2012.
11. Ignatyev, I.A., Thielemans, W. and Vander Beke, B. Recycling of polymers. A review. *ChemSusChem* 7(6), 1579–1593, 2014.
12. Scheirs, J. *Polymer Recycling. Science, Technology and Applications*. New York: Wiley, Chichester, 2001.
13. Panda, A.K., Singh, R.K. and Mishra, D.K. Thermolysis of waste plastics to liquid fuel: A suitable method for plastic waste management and manufacture of value added products—A world prospective. *Renewable and Sustainable Energy Reviews* 14(1), 233–248, 2010.
14. Al-Mosawi, A.I., Abdulsada, S.A. and Rijab, M.A. Mechanical properties of recycled bamboo fibers reinforced composite technical solutions. *European Journal of Advances in Engineering and Technology* 2(4), 20–22, 2015.
15. Ramzy, A. *Recycling Aspects of Natural Fiber Reinforced Polypropylene Composites* (Doctoral Thesis). Faculty of Natural and Materials Science, Clausthal University of Technology, 2018.
16. Yildirir E., Onwudili, J.A. and Williams, P.T. Recovery of carbon fibres and production of high quality fuel gas from the chemical recycling of carbon fibre reinforced plastic wastes. *Journal of Supercritical Fluids* 92, 107–114, 2014.
17. Iwaya, T., Tokuno, S., Sasaki, M., Goto, M. and Shibata, K. Recycling of fiber reinforced plastics using depolymerization by solvothermal reaction with catalyst. *Journal of Material Science* 43, 2452–2456, 2008.

18. Li J., Xu P.L., Zhu Y.K., Ding J.P., Xue, L.X. and Wang, Y.Z. A promising strategy for chemical recycling of carbon fiber/thermoset composites: Self-accelerating decomposition in a mild oxidative system. *Green Chemistry* 14, 3260–3263, 2012.
19. Torres, A., De Marco, I., Caballero, B.M., Laresgoiti, M.F., Legarreta, J.A., Cabrero, M.A., Gonzalez, A., Chomon, M.J. and Gondra, K. Recycling by pyrolysis of thermoset composites: Characteristics of the liquid and gases fuels obtained. *Fuel* 79, 897–902, 2000.
20. Cunliffe, A.M. and Williams, P.T. Characterization of products from the recycling of glass fibre reinforced polyester waste by pyrolysis. *Fuel* 82, 2223–2230, 2003.
21. Meyer, L.O. and Schulte, K. CFRP-recycling following a pyrolysis route: Process optimization and potentials. *Journal of Composite Materials* 43, 1121–1132, 2009.
22. Oliveira Nunes, A., Barna, R. and Soudais, Y. Recycling of carbon fiber reinforced thermoplastic resin waste by steam thermolysis: Thermo-gravimetric analysis and bench-scale studies. In *Proceedings of the 4th International Carbon Composites Conference (4th IC3), 12–14 May, Arcachon, France,* 2014.
23. Meyer, L.O., Schulte, K. and Grove-Nielsen, E. CFRP-recycling following a pyrolysis route: Process optimization and potentials. *Journal of Composite Materials* 43, 1121–1132, 2009.
24. Ogi, K., Nishikawa, T., Okano, Y. and Taketa, I. Mechanical properties of ABS resin reinforced with recycled CFRP. *Advanced Composite Materials* 16, 181–194, 2007.
25. https://www.materialstoday.com/composite-industry/features/recycling-composites-faqs/
26. Feih, S., Boiocchi, E., Mathys, G., Mathys, Z., Gibson, A.G. and Mouritz, A.P. Mechanical properties of thermally-treated and recycled glass fibres. *Composites Part B* 42, 350–358, 2011.
27. Cho, H.S., Moon, H.S., Kim, M., Nam, K. and Kim, J.Y. Biodegradability and biodegradation rate of poly (caprolactone)-starch blend and poly (butylene succinate) biodegradable polymer under aerobic and anaerobic environment. *Waste Management* 31(3), 475–480, 2011.
28. Mülhaupt, R. Green polymer chemistry and bio-based plastics. Dreams and reality. *Macromolecular Chemistry and Physics* 214(2), 159–174, 2013.
29. Davis, G.U. Open windrow composting of polymers. An investigation into the operational issues of composting polyethylene (PE). *Waste Management* 25(4), 401–407, 2005.
30. Tokiwa, Y., Calabia, B.P., Ugwu, C.U. and Aliba, S. Biodegradability of plastics. *International Journal of Molecular Sciences* 10(9), 3722–3742, 2009.
31. Bai, Y., Wang, Z. and Feng, L. Chemical recycling of carbon fibers reinforced epoxy resin composites in oxygen in supercritical water. *Materials and Design* 31, 999–1002, 2010.
32. Pimenta, S. and Pinho, S.T. The effect of recycling on the mechanical response of carbon fibres and their composites. *Composite Structure* 94, 3669–3684, 2012.
33. Al-Salem, S.M. and Lettieri, P. Kinetics of polyethylene terephthalate (PET) and polystyrene (PS) dynamic Pyrolysis. *World Academy of Science, Engineering and Technology* 4, 402–410, 2010.
34. Baillie, C., Matovic, D., Thamae, T. and Vaja, S. Waste-based composites—Poverty reducing solutions to environmental problems. *Resources, Conservation and Recycling* 55(11), 973–978, 2011.
35. Hadi, A.J., Najmuldeen, G.F. and Ahmed, I. Polyolefins waste materials reconditioning using dissolution/reprecipitation method. *APCBEE Procedia* 3, 281–286, 2012.
36. Hopewell, J., Dvorak, R. and Kosior, E. Plastics recycling. Challenges and opportunities. *Philosophical Transactions of the Royal Society B: Biological Sciences* 364(1526), 2115–2126, 2009.
37. Kartalis, C.N., Papaspyrides, C.D. and Pfaendner, R. Closed-loop recycling of post used PP-filled garden chairs using the restabilization technique. III. Influence of artificial weathering. *Journal of Applied Polymer Science* 89(5), 1311–1318, 2003.

Index

D

Damping factor, 87–88
Defects, 73
Delamination growth, 34
Delaunay meshing algorithm, 34
Design of experiments (DoE)
    based on factors settings, 25–27
    based on outputs, 25–28
    conceptual design, 27–29
    samples design, 29–30
Dilo, 53
DNA Hoechst method, 81
DoE, *see* Design of experiments

E

Eco-Compass, 96
Eco-composites, 107
Electrical resistance, 78
Electrostatic adhesion, 16
Elongation measurement, 77
Energy curve, 64
Energy release rate, 64
EN ISO 6946, 76
Environmental scanning electron micrographs
    (ESEM), 63
Epoxy resin, 96
ESEM, *see* Environmental scanning
    electron micrographs
Extensometer, 77

F

Fatigue coefficients, 64
FEA, *see* Finite element analysis
Feedstock/tertiary recycling, 118
Fibre blooming, 90
Fibre composite weakening, 64
Fibre matrix adhesion, 1
Fibre reinforced plastic (FRP), 47
Fibre reinforced polymer (FRP) composites,
    90, 97
Fibre reinforcements
    compression moulding, 51
    filament winding, 49–50
    lamination, 49
    pultrusion, 50
    spray method, 48–49
    vacuum bagging, 51
    vacuum infusion, 51–52
Fibre tortuosity, 33
Filament winding process, 49–50
Finite element analysis (FEA), 33, 35–36
Flax fibre, 6, 16
Frequency-modulated thermal wave imaging,
    74–75

FRP, *see* Fibre reinforced plastic; Fibre
    reinforced polymer composites
Functionality, 117
Fungal retting, 82
Furniture making process, 110–112

G

Gasification, 118
Gaussian elimination technique, 36
Gel coats, 90
Global warming potential (GWP), 62
Gomuti/Arenga pinnata fibre, 6
Graph theoretic approach, 73
Green composites, 47
GWP, *see* Global warming potential

H

Halpin-Tsai equation, 35
Hand lay-up process, 49, 109
Hard model, 33
HCP, *see* Hollow conjugated polyester
HDPE, *see* High-density polyethylene
Hemicellulose, 14
High-density polyethylene (HDPE), 85, 108
High temperature and pressure (HTP)
    solvolysis, 119
Hollow conjugated polyester (HCP), 37, 53, 54
Honey-bee silk fibre, 3
Hot calendaring, 77
HTP, *see* High temperature and
    pressure solvolysis
Hydrocracking, 118
Hydrolysis process, 118, 119

I

Impact assessments, 61
Infrared (IR) thermography, 73
Insulation behavior, 88–89
Interfacial adhesion, 14
Interfacial bonding mechanism, 16–17
Interfacial compatibility, 13
    blends within natural fibres, 13–15
    blends with synthetic fibres, 15–16
        interfacial bonding mechanism, 16–17
Interfacial debonding, 34
Inventory investigation, 61
In vitro testing, 81–82
Isotactic polypropylene (i-PP), 85

J

Jute fibre, 16, 37, 53, 65
Jute-reinforced polymer composite, 102

Milton Keynes UK
Ingram Content Group UK Ltd.
UKHW040051071024
449327UK00019B/480